Biotech Challenges

Catherine Regnault-Roger

Biotech Challenges

 Springer

Catherine Regnault-Roger
University of Pau et Pays de l'Adour
(UPPA - E2S IPREM-CNRS)
Pau, France

ISBN 978-3-031-38236-9   ISBN 978-3-031-38237-6   (eBook)
https://doi.org/10.1007/978-3-031-38237-6

The translation was done with the help of an artificial intelligence machine translation tool. A subsequent human revision was done primarily in terms of content.

Translation from the French language edition: "Enjeux biotechnologiques, Des OGM à l'édition du génome" by Catherine Regnault-Roger, © Presses de MINES - TRANSVALOR, 60, boulevard Saint-Michel - 75272 Paris Cedex 06 - France 2022. Published by Presses de MINES - TRANSVALOR. All Rights Reserved.

This Springer imprint is published by the registered company Springer Nature Switzerland AG
The registered company address is: Gewerbestrasse 11, 6330 Cham, Switzerland

Paper in this product is recyclable.

This book is dedicated to all who promote biotechnological improvements against the mermaids of obscurantism during the last 50 years. Let's keep digging the furrow of progress and innovation!

Foreword for English Edition

I am honored and very happy to present this historical account "From GMOs to genome editing" by Catherine Regnault-Roger, which describes so finely the 50-year regulatory saga around the engineering of plant genomes.

This book comes at a very special time. It is now 50 years since the cloning of genes became possible and 50 years since we discovered a way to include plants among the "modifiable" organisms. We have demonstrated that in nature, there are bacteria that have developed this technology during evolution. We were therefore convinced that it would be a revolutionary technology, easily applicable to make new plants.

However, society resisted this, and today, when there is such a need for more sustainable, high yielding and innovative agriculture, the technology remains strictly limited.

In this foreword, I try to explain why we should use this "misunderstanding" as an opportunity to create better trust in science. Yes, plant scientists, be they agronomists, ecologists, or molecular physiologists, are calling for "Enlightenment for plants too."

"The Discourse on Method" by René Descartes guided us and brought us an understanding of the non-living world. We have been and are encouraged to reflect. But without providing much information on what thought is. Everyday life had taught us, for hundreds of millennia, what rational thought was, without most humans realizing it.

Twentieth-century studies of animal behavior have taught us about the progressive development of neurobiology and brain function throughout animal evolution. Plants could survive and thrive, using simple chemical signaling, like the pathways used by "early microorganisms," for billions of years. Through science, we have observed the molecular basis of the differences and analogies of different life forms. But everyday life is an ongoing attempt to leverage this knowledge to better manage our living conditions.

This is not the place to reflect further on why this "rationality" helped hominids grow so numerous that most infectious agents (bacteria, viruses, etc.) could barely control population growth. Or why simple mathematics led to sophisticated tools and technologies, enabling the mass production and trade of goods, with the result that our community life changed dramatically. But most of us can agree that increasing the population from a few million to 10 billion requires a huge production of goods, incompatible for a planet with limited resources. These are the facts. We must keep this in mind.

For scholars trained in the human arts and the social, political, and economic sciences, the first, simplistic reaction may be: overconsumption and unnecessary goods are responsible and must be contained. Or worse, these technologies that we have developed are responsible. Some "radical environmentalists" are proposing to constrain "some" technologies or at least start by not allowing new ones. "Let's first wait to understand why scientists say they are so important." The next step is to blame science. Science is the source of all the new technologies that reach us. Science is overturning our traditions. Finally, thinking is dangerous. I believe Rodin was concerned about this when he put "The Thinker" in his monumental sculptural work "The Gates of Hell."

It is a fact that the life sciences have been difficult to analyze rationally. The understanding of the chemistry of living organisms is extremely recent. It seems that all or at least the vast majority of chemical reactions in the living organism are oscillating and ready to reverse. Not illogical since "life" is a biological process interacting with other members of living communities and with the physical environment.

In the middle of the twentieth century, it became possible to study the interactions between soil microorganisms and plants, thanks to the development of new technologies in biological chemistry. The new knowledge has been used to improve soil fertility and plant health. Plant protection and increased food and feed production were necessary and successful. But in the meantime, the world's population has tripled.

With the identification of the molecular mechanisms underlying plant growth, development, and adaptability to environmental changes (weather, soil conditions, competition, and pathogens), plant research has entered a new era. Huge advances in DNA sequencing over the past decade have made it possible to study the biodiversity and functioning of soil microorganisms. Soil engineering is ready to start. Reforestation, or the fixation of CO_2 in wood, a matter of good use of solar energy, could begin very soon with genetically improved trees. Even in weak or worn ground. Enough dew can be produced from the mountain forest to provide the water needed for agriculture (see Peru and Ecuador). We would need a thorough textbook to cover the possibilities of genetic manipulation and the impact of the resulting innovations for society.

Being aware of all the new plants that have been built, tested in greenhouses, but never allowed to be grown in the field or marketed, is very instructive. But the legislation around the "precautionary principle" makes all this impossible. That's what we can conclude from this book.

Most "lay people" take lightly of their ability to understand and decide on matters concerning a specialized science. We see that information technology and artificial intelligence are now fundamental to data processing. However, a significant portion of highly qualified experts in the humanities and arts believe that it is preferable to stop utilizing AI. Does this imply that a human authority might once more halt science devoid of any scientific evidence?

Many of us are inclined to believe that a statement like the one used against GMOs – "it's not because you can do it, you have to" – can be correct. It can be fine if we see serious negative points. If, however, we concur that nothing unfavorable can be demonstrated, but we don't allow it because someone claims we can't disprove it is potentially dangerous, then we have a major conflict in our assessment of science. If, above all, we have arguments that the genetic engineering of plants can make a substantial contribution to alleviating the many major environmental problems, then we can ask for your highest and most sustained attention on how decisions were made by the political authorities in Germany, France, and finally throughout the EU. This is the central message of this book, and it is of crucial importance for science and the future of our planet.

Finally, I would like to stress that communicating a serious societal problem can only be done after a long and in-depth process of reflection. After reading about the procrastinations in GMO approvals, I hope you will come to the following conclusion: Dare to think but watch what you think.

Marc Van Montagu

Foreword for French Edition

Catherine Regnault-Roger's project to write a book on the contribution of biotechnologies to the agricultural revolution was ambitious. The result lives up to expectations, as she has shown perfectly how, from GMOs (genetically modified organisms) to products derived from *New Genomic Techniques* (NGT), we have gone from "sledgehammers to molecular scalpels," in the words of Jennifer Doudna, co-Nobel Prize winner for Chemistry in 2020.

Innovation has always been the driving force behind the agricultural revolution. For nearly 10,000 years, humans have been selecting plant varieties, harvest after harvest, to improve their properties. The example of corn is enlightening. Discovered more than 7000 years ago in the highlands of Mexico, teosinte had a branched stalk, measured 2.5 cm, and had only a few grains per ear. It was already 10 cm long at the beginning of the Christian era. After its introduction in Europe in 1494, corn is today very compact, and its ear measures 30 cm. Man has always been able to select and improve varieties.

Catherine Regnault-Roger has the will to transmit her knowledge; she shows in a very pedagogical way that, by a curious return of history, the new genomic techniques make it possible to carry out genetic modifications that could have occurred naturally, given the natural variability of living organisms. These new techniques represent a fundamental break with the "old" GMOs, insofar as the modifications of the genome are much more targeted, safer, more precise, and faster. Man is only accelerating the course of events. The time scale, like the field of possibilities, is changing. Moreover, the traits sought are no longer just about resistance to herbicides, increased tolerance to diseases or insects, or improved productivity as with GMOs, but will allow the creation of new plants that consume fewer inputs, especially fertilizers, make better use of nitrogen, improve yields and quality, extend the duration of consumption, increase tolerance to diseases, adapt better to climate change, and select varieties that are more water-efficient, more resistant to water stress or salinity. As the author writes, it is a new era that is coming, not only in agriculture, but also in health, energy, and sustainable development.

Catherine Regnault-Roger not only has a recognized expertise in biotechnology research, but the fact that she is a member of the French Academy of Agriculture and has been a member of the High Council for Biotechnology gives her a special competence in the legal analysis of varieties obtained through these new technologies and in the international comparisons of these products. This is an obviously innovative and major chapter of a part of this book.

One would have thought that political decision-makers would be delighted with these new avenues allowing France to be more innovative and to reconcile agroecology and biotechnology. This has not been the case, as the governments concerned for the past 20 years have ignored the issue, leaving the decisions to be made to the judges.

Today, the fight that initially concerned transgenesis has shifted to all varieties obtained by mutagenesis. By speaking of *"hidden GMOs"* for varieties created using random or targeted mutagenesis techniques, the systematic opponents want to achieve the same effects and are betting on a new regulatory deadlock. They are about to succeed with the recent decisions of the Court of Justice of the European Union (CJEU) in 2018 and the French Council of State (*Conseil d'État*) in 2020, which are, for the vast majority of experts, scientifically aberrant and legally questionable, especially because the European judge did not rely on the scientific knowledge of the 2020s, but on that of the years when Directive 2001/18 was issued, when targeted genome modification techniques were not known. The latter exempted from the constraints imposed on genetically modified plants only those products "whose safety has long been proven." For the judge, whatever is the oldest is the safest. This doctrine slows down innovation, goes against the successive contributions of technologies, and calls into question the very notion of progress. One can imagine the brakes on progress in medicine if we had proceeded in this way.

A worrying conclusion of the author is that because of these controversies, France is gradually losing its capacity for international expertise in the field of plant biotechnology. This creates a gap between the European position and that of the rest of the world.

Finally, it addresses a major issue that of consumer perception, which remains far removed from the contribution that innovation could make.

The successive controversies, the shelling, and the activism of certain organized groups have struck the public opinion on health risks which, today, with 20 years of hindsight, are however not proven. In fact, the opposite is true: NGTs make it possible to use environmentally friendly industrial processes and to fight global warming more effectively.

We had written in 2017 in a resolution entitled *Sciences and Progress in the Republic*, unanimously adopted by the French National Assembly, that there is today a confusion, voluntarily maintained, between knowledge and beliefs, which threatens the foundations of scientific research and of our democracy. While the subjects are complex, the explanations are increasingly simplistic. The traditional mechanisms of sorting information are abolished. Everything is mixed up: validated knowledge, fanciful hypotheses, sectarian speeches, and conspiracy theories. The premium is given to striking, and therefore extreme, information. Accurate and nuanced information is often less attractive. Hence the proliferation of ideologies, which feed on a mixture of blindness, propaganda, and random knowledge.

However, this mistrust undoubtedly has a deeper cause. Since the end of the 2000s, repeated disasters such as the Chernobyl and Fukushima nuclear accidents, the contaminated blood scandal, and then mad cow disease, combined with the growing perception of the effects of climate change, have convinced our fellow citizens that technological progress does not go hand in hand with human progress. The philosopher Claude Debru declared in a symposium organized by the Academy of Agriculture in 2020: "without innovation, there is no progress." But, in return, progress must, of course, be both controlled and shared.

As the author says, the fundamental question is how, in the second half of the twenty-first century, we can feed nearly 10 billion people without further draining and destroying the planet. To achieve this, we must reconcile biotechnology and agro-ecology. If old Europe does not consider innovation in agriculture as a priority, it is heading for decline. Catherine Regnault-Roger pedagogically analyses the differences between risks and dangers. Risks should not be dismissed out of hand, but rather evaluated rationally, keeping beliefs, ideological biases, propaganda, and sectarian discourses at bay.

This book is fascinating, well argued, and does not evade any aspect of this question of the potential contribution of biotechnologies to agriculture. It shows that the author has convictions, but that these are based on knowledge. It popularizes a complex subject and opens perspectives for the future. It will become a reference for students, researchers, lawyers, and industrialists and honors the work of a member of the French Academy of Agriculture.

Jean-Yves Le Deaut

Preface

The book you have in your hands aims to bring together some of the knowledge acquired in the field of biotechnologies that implement genomic modifications, starting with considerations on the applications carried out and envisaged, as well as the regulations applied to the products obtained and the geopolitical consequences that they induce.

At the end of a mandate of more than 12 years as a scientific expert in ecotoxicology within the Scientific Committee of the High Council for Biotechnology (from its creation in 2009 until the end of its existence in December 2021), I felt the need to draw up a report on the current stakes in terms of biotechnological innovations and above all to rectify preconceived ideas and misinformation on this controversial societal subject.

This book was updated in the spring of 2023 from a book written in 2021 in French and published by Presses des Mines in the collection *Académie agriculture de France* in May 2022. I thank Editions Springer-Nature, and especially Ms. Zuzana Bernhart, executive editor, for having accepted to publish this English version.

The French language edition has been foreworded by Jean-Yves Le Déaut, member emeritus of the French Academy of Agriculture and honorary parliamentarian member after a long parliamentary career as a Socialist Party deputy in the National Assembly of the French Republic for 32 years. Known for his scientific commitments, organizer of the first citizen's conference on GMOs (Genetically Modified Organisms) in 1998, Jean-Yves Le Déaut is particularly well placed to present this book. President for many years of the Parliamentary Office for the Evaluation of Scientific and Technological Choices (OPECST), he is a particularly knowledgeable observer, especially since in his previous life he was a university professor of biochemistry and director of a laboratory at the University of Lorraine. The numerous parliamentary reports of the OPECST that he has produced on difficult subjects such as chlordecone or what has become the *Séralini affair* testify to his willingness to present with rigor thorny subjects in all their facets without hiding the difficulties. I deeply thank him for the honor of forewording this book.

Professor Marc Van Montagu has agreed to write the preface of the English version of this book and I am not only honored but also very moved. Indeed, who does not know this great pioneer scientist who opened the way to modern biotechnology?

He is the father of transgenesis, winner with Mary Dell Chilton and Robert Fraley of the prestigious *World Food Prize Foundation* in 2013 and of the no less remarkable *Japan Prize* with Jozef Schell, with whom he developed the first transgenic tobacco that secretes insecticidal proteins to defend itself against its pests by gene transfer using an agrobacterium. His illustrious work has earned him worldwide recognition. For more than 30 years, he has been a member, now emeritus, of the French Academy of Agriculture (elected in 1992 as a foreign member) and a founding member of the French Association of Plant Biotechnology (AFBV) created in 2009. Beyond the many honors he has received, Marc Van Montagu continues to actively support the work and development of biotechnologies in their most modern perspectives. It is therefore a great honor for me that he has accepted to foreword this book devoted to "Biotechnology Challenges" subtitled "From GMOs to Genome Editing."

Initially published, in the *Académie d'agriculture de France* collection of Presses des Mines, this book was supported by the Book Committee of the Academy with the assistance of two academicians whose wise advice supported the editorial process. I warmly thank Professor emeritus Nadine Vivier, honorary president of the Academy, and Doctor Léon Guéguen, emeritus member of the Academy, for their attentive eyes and their friendly suggestions. I also thank Dr. Sergio Ochatt, member of Scientific Committee of High Council of Biotechnologies for his friendly advises regarding this English edition.

I would also like to express my gratitude to those who encouraged the distribution of the French edition, in particular the academicians Michel Thibier, emeritus member of the French Academy of Agriculture and honorary president of the French Veterinary Academy and of the European Union of Agricultural Academies (UEAA), and François Desprez, member of the Academy of Agriculture who also has eminent responsibilities in the seed interprofession.

Finally, I would like to thank all those who participated in the success of the book by organizing breakfast and diner debates (Institut Sapiens, Association des Amis de l'Académie d'agriculture de France, LifeHub Lyon and many local associations of Basque Country: SMLH Côte basque, Rotary-BAB, UTL Biarritz et UTL Saint-Jean-de Luz, where I live), or with interviews and press articles published in *La France Agricole*, *The European scientist*, *IREF-Europe*, *Phytoma-ldv*, *la République des Pyrénées*, *Genetic Literacy Project* (GLP), *l'Opinion*, *Paysans et Sociétés*. And I apologize to those I have forgotten.

Through this book, I am pleased to offer you an essay that examines the recent past but also the present. It leads to a well-founded reflection on a subject positioned at the heart of innovation and the agri-food independence of countries, therefore on a subject that is decisive for our future.

Biarritz, France   Catherine Regnault-Roger

Contents

Part IV   General Conclusion

Chapter 1
Introduction

The Covid-19 pandemic has disrupted world trade during the year 2020, the movement of goods and people, and in so doing, has highlighted how essential it is for a country to have productive autonomy to provide for the needs of its population in such vital sectors as agri-food, health, or energy.

The international division of labor, accompanied by the relocation of production to countries where the cost of labor is more attractive, has led to shortages of products that are essential for human's health, such as medicines, medical equipment, and masks. Because they did not have the necessary industrial facilities on their territory, some countries, including France, faced a health emergency, not without difficulties, at the beginning of the pandemic. Fortunately, the situation was better at the agricultural level, and European production made it possible to feed the population. However, let's be careful not to be triumphalist, as the *French Cour des Comptes* (Court of Auditors) denounced in its March 5, 2019, report a situation that it considers alarming for French agriculture, which is steadily losing market share, dropping from the second place in the world in the 1990s to the fifth place now [1]. Although agricultural exports have held up better in the context of 2020 than those of the aeronautics sector, they are still down by 3% compared to the previous year, according to FranceAgriMer [2]. It is in this tense context that the question arises as to the place that biotechnological innovations should occupy and the regulations that are reserved for them in the European framework that is now theirs.

Agricultural biotechnologies are in the front line because of a societal distrust that is manifesting itself in European countries. Public opinion campaigns and illegal media exactions orchestrated by militant NGOs (non-governmental organizations), marked by a politically oriented ideology of degrowth, have instilled doubt about the benefits of genetic progress, more particularly about biotechnologies applied to the varietal improvement of cultivated plants and to breeding, the human health sector being less subject to this orchestrated disinformation until now.

The GMOs (genetically modified organisms) generated by certain biotechnological techniques applied to plants and livestock are still the subject of intense debate, and that of products derived from New *Genomic Techniques* (NGTs) is now at the heart of this debate.

The characteristics of certain products obtained by NGTs make it difficult to identify the modifications carried out in the laboratory: these can also occur spontaneously in nature or be the result of classic techniques used for over a hundred years. Apart from a voluntary declaration of the production process, this lack of distinction makes it difficult to trace these products. In these conditions, is it the process of obtaining that is important or are the characteristics of the finished product which matter?

Many countries have already taken note of these technological advances by adapting their regulations applied to products derived from these new genomic techniques, while the Court of Justice of the European Union (CJEU) ruled in a judgment of July 25, 2018, that products obtained by biotechnological techniques developed after 2001, regardless of the modifications made, must be subject to the regulations applied to GMOs in the European Union (EU), i.e., Directive 2001/18/EC. This position is unique in the Northern Hemisphere.

New Zealand, which had a similar ruling from the High Court of Justice in 2014, is now considering some changes may be possible for "boosting productivity, improving health outcomes, reducing biosecurity risks, and responding to climate-change risks and other environmental problems effectively and efficiently," according to the report 2021 of the highly official *New Zealand Productivity Commission to the government* [3].

Can the European Union (EU) alone adopt rules that isolate it from the rest of the world? Aware of the situation, the European Commission published a document at the end of April 2021, opening a debate within the EU and giving rise to a wide-ranging consultation open to all: economic players, various organizations, and ordinary European citizens. This consultation is intended to review the "*Legislation for plants produced by certain new genomic techniques*" according to the title given to the current process [4].

In this book, we propose to provide some information and food for thought in order to clarify the context of the development of biotechnologies, their applications and the global situation, as well as the regulatory and geopolitical issues.

The book is divided into three parts. The first part will give some background on biotechnologies, their definition, and will question the regulations applied to them, while the second part will draw up an assessment of GMOs in their medical and agricultural applications, and the third part will present the prospects opened by NGTs and the state of research in the world.

Bibliography

1. Cour des Comptes. (2019). *Les soutiens publics nationaux aux exportations agricoles et agro-alimentaires*, Référé n°S2019-0467 du 5 mars 2019, https://www.ccomptes.fr/system/files/2019-05/20190520-refe-S201

2. FranceAgriMer. (2021). Performances à l'export des filières agricoles et agroalimentaires françaises en 2021 (The export performance of the agricultural and agri-food sectors in 2020) seminar of May 17, 2021, https://www.franceagrimer.fr/Actualite/Etablissement/2021/Les-performances-a-l'-export-des-filieres-agricoles-et-agroalimentaires-en-2020
3. Office of the Prime Minister's Chief Science Advisor. (2022). *Gene editing*, https://www.pmcsa.ac.nz/topics/gene-editing/
4. European Commission. (2021). *Legislation for plants produced by certain new genomic techniques*, https://ec.europa.eu/info/law/better-regulation/have-your-say/initiatives/13119-Legislation-for-plants-produced-by-certain-new-genomic-techniques_en

Part I
Biotechnologies: Benchmarks and Regulatory Issues

Chapter 2
Biotechnology: Timeless, Essential, and Omnipresent

Biotechnologies were initially implemented empirically during prehistory, probably as early as the Neolithic period, when the life of the human species was radically changed with the advent of agriculture, and then deliberately, thanks to the progress in knowledge and understanding of the observed phenomena that accompany the history of mankind. The ancient Egypt of the Pharaohs already practiced the manufacture of fermented foods (beer, cheese, etc.) and domesticated animals, as evidenced by papyrus and frescoes found on temples and pyramids. Biotechnologies, from "protobiotechnologies"[1] (as coined by Karl Ereky in 1913 [1]) to modern biotechnologies, have accompanied the epic of human development and are, still today, indispensable tools for innovating and proposing solutions for the challenges of the future.

A Broad Definition

The definition of biotechnology used today is that of the OECD[2], a definition that is deliberately broad in order to cover "all modern biotechnology, but also many traditional activities" [2], but which is accompanied, for the sake of clarification, by a list that also, deliberately, is very precise and technical, but which may evolve according to advances in knowledge, application targets, and the methods concerned (Box 2.1). The aim of biotechnology is to manufacture products from raw materials using living organisms, as summarized by the science journalist Nathalie Meyer:

---

[1] The protobiotechnologies are situated from the Neolithic-Antiquity until the beginning of the twentieth century.

[2] Organisation for Economic Co-operation and Development. This international organization is 38 countries which share a democratic system of government and a market economy.

"biotechnology is a subtle marriage between the science of living beings and technology" [3].

The late Axel Kahn, who died in 2021 and was a fervent defender of genetic progress, gave a similarly broad definition in 1996 based on the concept of "the use of living cells to produce various substances" [4], a definition for which he specified that it was necessary to distinguish between biotechnologies taken as a whole (the results of various interactions) and genetic engineering, which develops a direct manipulation of the genome of organisms.

Box 2.1: OECD Definition of the Concept of Biotechnologies and Associated Techniques
Definition

"The application of science and technology to living organisms, as well as parts, products and models thereof, to alter living or non-living materials for the production of knowledge, goods and services.

The list-based definition of biotechnology techniques

DNA/RNA: Genomics, pharmacogenomics, gene probes, genetic engineering, DNA/RNA sequencing/ synthesis/amplification, gene expression profiling, and use of antisense technology.

Proteins and other molecules: Sequencing/synthesis/engineering of proteins and peptides (including large molecule hormones); improved delivery methods for large molecule drugs; proteomics, protein isolation and purification, signaling, identification of cell receptors.

Cell and tissue culture and engineering: Cell/tissue culture, tissue engineering (including tissue scaffolds and biomedical engineering), cellular fusion, vaccine/immune stimulants, embryo manipulation.

Process biotechnology techniques: Fermentation using bioreactors, bioprocessing, bioleaching, biopulping, biobleaching, biodesulphurisation, bioremediation, biofiltration and phytoremediation.

Gene and RNA vectors: Gene therapy, viral vectors.

Bioinformatics: Construction of databases on genomes, protein sequences; modelling complex biological processes, including systems biology.

Nanobiotechnology: Applies the tools and processes of nano/microfabrication to build devices for studying biosystems and applications in drug delivery, diagnostics, etc."

Wide Fields of Application

Biotechnologies today have many fields of application, all of which are colored: marine (blue biotechnologies), plant (green), health (red), industrial (white), environmental protection (yellow biotechnologies), or learning (orange). They have become the cornerstone of the bioeconomy, which is set to grow in the coming decades. Their implementation is based on multiple approaches combining genome modification, in vitro tissue culture, bioreactors, nanobiotechnology processes, and computer modeling applied to biological transformations. They have become essential to current and future developments.

The biotechnologies provide service in the medical field, where they have enabled spectacular advances in the prevention or treatment of diseases with vaccines or drugs (e.g., the biosynthesis of human hormones such as insulin or growth hormone by genetically modified bacteria), or new therapeutic approaches, or in the environmental sphere, with the use of microorganisms or plants to eliminate waste and clean up soils or to develop biocontrol solutions[3] to limit the use of synthetic plant protection products, and also in agriculture and food, with the development of new, more efficient plant varieties adapted to the environment or foodstuffs that are better suited to food safety and consumer tastes. Biotechnologies are today an essential tool, just as they were in the past, when our ancestors produced the first wines, the first yoghurts, or even the sauerkraut, vinegar, and bread in an empirical way.

This wide range of food products is the result of anaerobic and aerobic fermentations (without and with oxygen) in the presence of microorganisms that synthesize lactic acid, acetic acid, or ethanol and give their organoleptic and nutritional properties to food. We consume every day biotechnological food products, even if it is only "our daily bread[4]"!

Thus, biotechnologies have always been part of humans' environment and have helped him to live better. People have used it without being aware of it, like Mr. Jourdain (the famous character in Molière's *Le Bourgeois gentilhomme*) spoke in prose without knowing it.

The Rise of Biotechnology Through Genetic Engineering

With the discovery of the structure of DNA (deoxyribonucleic acid) by Jim Watson and Francis Crick in 1953, which was accompanied by the discovery of RNA in 1961 by Jacques Monod, François Jacob, and François Gros, and then the mechanisms of protein synthesis with the deciphering of the genetic code in the years that

---

[3] For example, the use of antagonistic biological organisms (biological control) or compounds extracted from microbiological or plant organisms [5].

[4] Well-known Christian prayer (*Pater noster*).

followed by Marshall Nirenberg and Har Gobind Khorana in 1966 [6], the most spectacular advances in biotechnology have made it possible to identify the functional cellular mechanisms and the consequences of spontaneous or voluntary genome modifications.

The first recombinant DNA[5] and then the first genetically engineered bacterium with the lambda phage[6] were produced by Paul Berg in 1972 [7]. Other teams using different vectors (plasmids)[7] to transfer foreign DNA to the recipient species contributed to the development of these first genetically modified microorganisms [8]. This was followed by work on more complex, multicellular organisms. In 1981, the first transgenic mouse was created for laboratory studies, and in January 1983, the first transgenic plants, tobacco and petunias, were presented at a conference in Miami (see Chap. 11). GMOs (genetically modified organisms) were becoming tangible. These innovations opened the door to applications in both medical and agricultural fields.

Thus, as early as the 1980s, bacteria were genetically modified to produce therapeutic compounds in bioreactors, such as recombinant human insulin to treat insulin-resistant diabetes (1982) or recombinant growth hormone obtained by genetic engineering (1986) after deaths were observed in the United States in 1984 with growth hormone treatments extracted from human cadavers of uncertain health history.

Transgenic animals, not only mice but also rabbits, sheep, and pigs, were also genetically modified by direct microinjections for therapeutic or veterinary research purposes and to improve animal husbandry in 1985 [9]. Three years later, in 1988, the *OncoMouse®* was patented by Harvard University (MA, USA) and commercialized by its economic partner, the DuPont de Nemours company, for cancer studies. This commercialization of a "nonhuman transgenic mammal" (the name of the *OncoMouse®* in the patent) caused a controversy that lasted 12 years. In 2000, an agreement was reached between the NIH (*National Institutes of Health)* and the DuPont company to withdraw the animal from the market but to allow university and federal researchers to continue to use it for their work [10].

Genetically modified plants (GMP) are not to be outdone. A tobacco [11] and then a tomato [12] were genetically modified to resist insects by bacterial transfer using *Agrobacterium tumefaciens* in 1987. The latter is "disarmed," i.e., deprived of its infectious gene that spreads crown gall, a plant root disease. It was replaced by the gene from *Bacillus thuringiensis*, a widespread soil bacterium that synthesizes the Cry 1Ab protein and whose toxic properties for many insects were shown in 1981 [13].

---

[5] DNA modified in the laboratory in vitro.

[6] Bacteriophage virus that infects the bacterium *Escherichia coli* and is used as a vector of genetic material.

[7] Small circular DNA fragments present in the bacterial cell and independent of the bacterial genome.

In the same year, 1987, the technique of the microparticle gun was described, which consists of the projection of microbeads coated with the genetic construction that is to be inserted into the genome of the recipient organism. This is direct transfer by biolistics [14] (see Chap. 4).

In the 1990s, the first genetically modified plants were marketed. In addition to virus-resistant tomatoes and tobacco plants grown in China [15, 16], a tomato with delayed ripening, the *Flavr Savr* tomato produced by the company Calgene, was marketed in the United States in 1994, but it was a failure (see Chap. 11). In 1996, this pioneering venture came to an end. That same year, the epic of the first transgenic crops in six countries began (see Chap. 12).

At that time, many genetic modification projects on several plant species existed in the boxes of European and American research laboratories[8], as shown by the patents taken. Thus, between 1989 and 1994, American companies filed 176 patents concerning plant biotechnologies, while 203 were filed by European companies. This was a time when newspapers could headline "When the European authorities lead the way" [17]! The public sector expenditure of the EEC[9] was equal to that of the United States (240 million Francs) and represented 44% of world expenditure on research and development (R&D) (i.e., 88% for the two economic powers that dominated the world at the time). European R&D competes unabashedly with American laboratories.

Genetic engineering was already opening huge prospects at that time:

- In the field of environment and agriculture: The cultivation of genetically modified plants concerns an improvement in the quality and quantity of harvests (fight against diseases, crop pests or weeds, by integrating resistance traits against insects, for example) and an improvement in technical itineraries with a reduction in phytosanitary inputs or better adaptability to difficult environments.
- In the food field: Improvement of taste and nutritional values, better preservation of the food.
- In the medical and pharmaceutical field: Treatments by gene therapy, creation of vaccines, production of therapeutic molecules or new drugs.

The field of possibilities is immense, and the achievements will be detailed in part 2 of the book.

But first we need to remove some concerns and counter the preconceived notions that these technological advances raise.

---

[8] Reference should be made to the numerous special reports in *La Recherche* or *Biofutur* in the 1990s, for example, La Recherche n° 270 (1994), *Biofutur* n° 90 (1990) or 172 (1997).

[9] The EEC (European Economic Community) in 1990 was composed of 12 countries (France, Germany, Belgium, the Netherlands, Luxembourg, Italy, the United Kingdom, Ireland, Spain, Portugal, Denmark, and Greece).

Bibliography

1. Fárl, M. G., & Kralovánszky, U. P. (2006). The founding father of biotechnology: Károly (Karl) Ereky. *International Journal of Horticultural Science, 12*(1), 9–12. https://doi.org/10.31421/ IJHS/12/1/615

2. OECD. (2019). *Statistical definition of biotechnology* (updated 2005), https://www.google.com/search?q=https%3A%2F%2Fwww.oecd.+org%2Fen%2Fsti% 2Femerging-technology%2Fstatistical-definition-of-biotechnology-updated-2005.+ Html&rlz=1C1GCEU_enIN1020IN1020&oq=https%3A%2F%2Fwww.oecd.+org% 2Fen%2Fsti%2Femerging-technology%2Fstatistical-definition-of-biotechnology-updated-2 005.+Html&aqs=chrome..69i57j69i58.624j0j7&sourceid=chrome&ie=UTF-8

3. Meyer, N. (2016). Biotechnologie qu'est ce que c'est? *Futura Tech*, https://www.futura-sciences.com/tech/definitions/technologie-biotechnologie-15588/

4. Khan, A. (1996). *Transgenic plants in agriculture* (p. 165). John Libbey EUrotext.

5. Regnault-Roger, C. (2014). *Produits de protection des plantes. innovation et sécurité pour une agriculture durable* (p. 368). Lavoisier.

6. Chevassus-au-Louis, N. (2021). Décryptage du code génétique, *Dictionnaire de l'Encyclopédie Universalis*, https://www.universalis.fr/encyclopedie/decryptage-du-code-genetique/

7. Jackson, D., Symons, R. H., & Berg, P. (1972). Biochemical method for inserting new genetic information into DNA of simian virus 40: Circular SV40 molecules containing lambda phages genes and the galactose operon of *Escherichia coli*. *Proceedings of the National Academy of Sciences of the United States of America, 69*, 2904–2909.

8. Morrow, J. F., Cohen, S. N., Chang, A. C. Y., Boyer, H. W., Goodmann, H. M., & Helling, B. (1974). Replication and transcription of eukaryotic DNA in *Escherichia coli*. *Proceedings of the National Academy of Sciences of the United States of America, 71*, 1743–1747.

9. Hammer, R. E., Pursel, V. G., Rexroad, C. E., Jr., Wall, R. J., Bolt, D. J., Ebert, K. M., Palmiter, R. D., & Brinster, R. L. (1985). Production of transgenic rabbits, sheep and pigs by microinjection. *Nature, 315*, 680–683.

10. Vogt, T. F. (2001). *OncoMouse® encyclopedia of genetics* (pp. 1372–1373). Academic Press.

11. Vaeck, M., Reynaerts, A., Höfte, H., Jansens, S., De Beuckeleer, M., Dean, C., Zabeau, M., Van Montagu, M., & Leemans, J. (1987). Transgenic plants protected from insect attack. *Nature, 328*, 33–37.

12. Fischhoff, D., Fischhoff, D. A., Bowdish, K. S., Perlak, F. J., Marrone, P. G., McCormick, S. M., Niedermeyer, J. G., Dean, D. A., Kusano-Kretzmer, K., Mayer, E. J., Rochester, D. E., Rogers, S. G., & Fraley, R. T. (1987). Insect tolerant transgenic tomato plants. *Nature Bio/Technology, 5*, 807–813.

13. Copping, L. G., & Hewitt, H. G. (Eds.). (1998). *Crops protection agents from nature and analogues* (p. 164). Royal Chemical Society UK.

14. Klein, M., Wolf, E. D., Wu, R., & Sanford, J. C. (1987). High-velocity microprojectiles for delivering nucleic acids into living cells. *Nature, 327*, 70–73.

15. Tao, Z., & Shundong, Z. (2003). L'utilisation des OGM en Chine: enjeux et débats. *Perspectives Chinoises, 76*, 52–60.

16. James, C. (1997). Global status of transgenic crops in 1997, *ISAAA Briefs* No. 5. ISAAA: Ithaca (NY), p. 31, https://www.isaaa.org/resources/publications/briefs/05/default.html

17. Hoeveler, A., & Magnien, E. (1997). Quand les instances européennes mènent le bal. *Biofutur, 172*, 12–15.

Chapter 3
Genomic Modification: The Essence of Biological Life

Concerns have been expressed about the development of biotechnologies that involve human-induced genome modification: in so doing, it has been raised; they could produce different, monstrous, and dangerous abnormal organisms, because this genomic modification is not "natural" but results from human intervention.

Incessant Genome Changes in Nature

It appears, however, that genome modification is a natural and constant phenomenon that does not need the human hand to occur. Laboratory techniques of genome modification, such as mutagenesis and transgenesis, which have been the subject of controversy over the artificiality of organisms obtained through biotechnology, occur spontaneously in ecosystems. Simply, before implementing them in these closed and controlled environments that are scientific laboratories, we did not know that these phenomena existed naturally because they were not characterized.

The modification of the genome of living beings is an essential phenomenon for the life of organisms. It allows them to adapt to face new interactions and environmental changes. These transformations, which make the necessary adjustments for the survival of species, generate a constant evolution of ecosystems and consequently of the biotope.[1] There is a great plasticity of the genome within all living species, which translates into the appearance or disappearance of very variable and varied characters. This allows organisms and their populations to cope with continuous changes in their environment. The evolution of species is based on these phenomena. Because of the wide range of responses made by the various organisms to these changes, it results in a biodiversity that constitutes the basis of the adaptation

---

[1] The biotope is a homogeneous biological environment suitable for the development of one or more species.

C. Regnault-Roger, *Biotech Challenges*,
https://doi.org/10.1007/978-3-031-38237-6_3

of living beings to respond to the continuous modifications of the biosphere.[2] And this is why it is important to maintain a level of biodiversity that allows this adaptability of species.

Understanding these phenomena has only been possible thanks to scientific work, including the transmission of genetic traits and biological heredity with the work of Gregor Mendel (1822–1884) rediscovered by Hugo de Vries at the beginning of the twentieth century and then the deciphering of the structure of DNA by Watson and Crick (cf. Chap. 2). These two major discoveries have shed light, on the one hand, on the functioning of hereditary genetic material and the links between genes and characteristics and, on the other hand, on the influence of the environment (climate, environment, and ecosystems).

Mutagenesis and Transgenesis: Spontaneous Phenomena

Research over the past decades has also shown that mutagenesis and transgenesis, techniques widely used to transform the genome, are both natural mechanisms.

Mutagenesis is the modification of the sequence of nitrogenous bases of the DNA double helix that occurs following the loss of one or more bases (deletion) or the replacement of a base by another base or a rearrangement of several bases. This results in a modification of the cellular genome and genetic characteristics. Thus, mutagenesis can be expressed by giving new properties to the cell and the organism, or it can go unnoticed when the changes are minimal: it is then a silent mutation that has affected the genome but does not confer, in the context of the moment, any tangible advantages or disadvantages. On the other hand, with a change in environmental conditions, the silent modification can be advantageous or disastrous. For the results of mutagenesis to last, it must confer an advantage on the mutant in its life and environment. A disadvantageous mutation results in the eventual disappearance of the mutant. Spontaneous mutation is therefore one of the motors of evolution, giving species the ability to adapt to changes in their environment and ecosystems. It produces individuals which, within a population, have the survival characteristics that allow them to develop and reproduce.

Transgenesis is also a natural phenomenon. It was not created by man. Whether viral or bacterial, DNA can naturally be incorporated into the genomes of eukaryotic organisms and genetically modify them. The resulting gene flow is accompanied by genetic recombination and is again one of the drivers of evolution. If this modification gives these organisms a selective advantage, they live; if not, they disappear in a more or less long period of time or are transformed again: crossbreeding and genetic mixing modify their performance. It has been found that the sweet potato *Ipomoea batatas* (L.) has acquired DNA fragments from soil bacteria of the

---

[2] The biosphere includes all living organisms and their living environments.

genus *Agrobacterium* during evolution and can be considered a GMO: it is a naturally transgenic organism.

Implementing mutagenesis and transgenesis techniques in the laboratory consists in provoking-oriented transformations of the genome, which are less random in their results than the phenomena observed spontaneously, and in acting on the time factor to program the acquisition of a character that can occur naturally.

Gene Transfer Between Species

Another concern that has been expressed about genetically engineered organisms is that the species barrier would be transgressed: by introducing into the genome of one organism genes from other species that are sexually incompatible, are we creating laboratory monsters? What is the reality of gene transfer between species in nature?

Recent research advances show that there is ample evidence that bacteria, plants, and animals have acquired genes by horizontal transfer: genes from fungi into insects and from algae into mollusks. Gene transfer between species that are not sexually compatible occurs spontaneously in nature, without the hand of man. It is therefore a natural phenomenon that can occur on a large scale and that two examples illustrate:

- The mitochondria of mammalian cells, which allow them to breathe and produce cellular energy from absorbed oxygen, are most probably the result of a symbiotic incorporation of bacteria, with an estimated contribution of 2000–3000 genes [1] in cells originally incapable of using oxygen. Evolution has subsequently sorted out the genes, with plants, for example, retaining more functional genes from this bacterial contribution than our species.
- Genome sequencing has made it possible to identify that plant genomes can be composed of several genomes of different origins in a previously unsuspected mosaic. For example, the wheat genome, whose sequencing has just been completed, contains 42 chromosomes and brings together 3 subgenomes of related species, each with 14 chromosomes that "interact with each other," according to Hervé Le Guyader [2]. Similarly, the rapeseed genome, whose sequencing was finalized in 2014, is formed by two subgenomes of plants crossed involuntarily, as it seems, by humans at the end of the Neolithic: cabbage and turnip.

Is the distinction between "natural" and "artificial" therefore justified? One of the latest genome-editing techniques that is currently developing very rapidly, CRISPR (see Chap. 4), causes targeted genomic modifications and makes it possible to obtain, under certain conditions, organisms that are indistinguishable as to whether they are the result of a genetic modification induced in the laboratory or of a spontaneous mutation present in the biotope and hitherto ignored, which occurred, for example, in an unknown species of the deep Amazon!

By transferring a gene from one species to another in order to repair a genetic damage or to develop a functionality linked to a new trait to allow an organism to resist a disease or to adapt to extreme conditions of drought, for example, at the most, we reuse a bolt from a car body to repair a washing machine, according to the delicious image evoked by Louis-Marie Houdebine [1]. What is expected of this bolt, regardless of its origin, whether it comes from a factory or a scrapyard, is that it should be in the right size and in the right condition to fulfill the function expected of it by restoring the washing machine's operation. But its presence within the mechanical parts of the washing machine will not change it into an Austin Mini or a Ferrari!

It is now known that an isolated gene does not have a character specifically linked to a species and that there are genes common to several species, for example, those that control common functions such as the energy metabolism of cells. Gene transfer between organisms exists spontaneously and is not exceptional in nature.

As it has been pointed out, "these few examples demonstrate that the crossing of the species barrier is a natural phenomenon and that the border between natural and artificial is nowadays very blurred" [3].

Genome Modification: An Ancestral Agricultural Practice

By abandoning gathering and hunting as a means of subsistence and developing agriculture and animal husbandry, man has empirically practiced genome modifications in the species he cultivated or raised. These agricultural and pastoral activities have been accompanied by the domestication of species selected to feed them, because, with agriculture, it is indeed a genetic selection that is carried out to have more abundant harvests and of better nutritional or sanitary quality or to raise herds that are more resistant to diseases and better producers of meat or milk. Thus, at first by empirical practices, then in a more reasoned and rational way with the progress of scientific knowledge, crossbreeding was carried out between plant and animal species that presented the best characteristics to meet the desired criteria. But before the significant advances in genetic and genomic knowledge, we did not know that by domesticating species, we were modifying their genomes.

The modification of the genome takes place through genetic mixing between species and sexually compatible varieties. Very early, since the highest antiquity, men have operated genetic mixing. The existence of the mule, a cross between the donkey and the horse to obtain an animal combining the strength of the horse and the robustness of the donkey, is an example.

In some cases, the domestication of species is accompanied by a considerable modification of the phenotype[3]. This is the case for the cereal that is now the most

---

[3]All the apparent characteristics of an individual.

widely cultivated in the world[4], maize (*Zea maize* species). A Central American plant, maize was the staple diet of the first Amerindian civilizations, which initially cultivated for its flour teosinte, a plant well adapted to the hot and humid climate. Generation after generation, the farmers select the plants that have the most beautiful ears, and the ears that have the most grains and yield the most flour, and also the plants that are most resistant to climate variations, rain, cold, or drought. They crossbreed individuals that have the best qualities. Gradually, the cultivated plant has become stronger in its environment and more productive. The plant's ears change morphology. From quasi-filiform and brown in teosinte, they thicken and become paunchy and yellow with a considerable number of bulging grains. The very educational illustration of the GNIS (Fig. 3.1) underlines this evolution.

Teosinte, now considered a subspecies of maize (*Zea mays* subsp. *parviglumis*), is genetically very close to maize but morphologically very different, with abundant tillering and small cobs that are easily shelled. The first corn is dated 7000 years BC and was in Central Mexico. It is estimated that at that time a corn cob measured about 2.5 cm and that yields were around 0.12 tons per hectare. As a result of human selection and spontaneous mutations, the adaptation to other environments gave various populations with different appearances of color or size of the ears. Today, the ears, according to the varieties, are from 5 to 45 cm with a diameter from 3 to 8 cm.

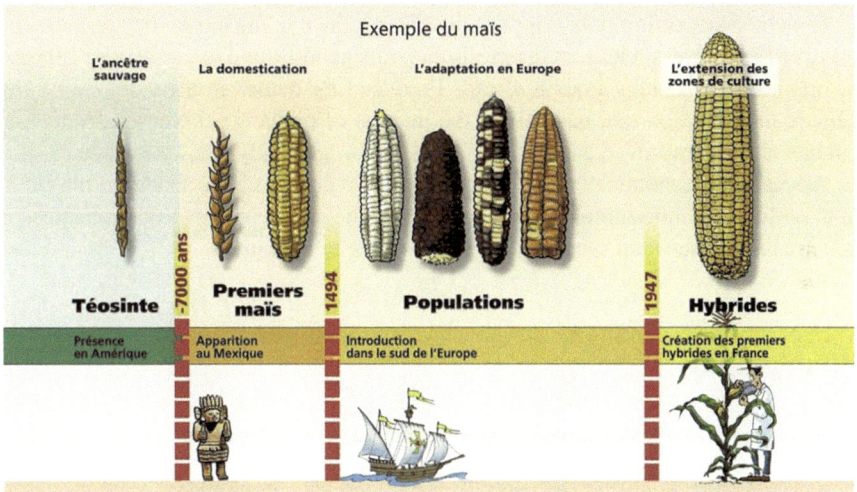

Fig. 3.1 Example of domestication of a crop: from teosinte to maize (Illustration GNIS: https://www.google.com/search?q=www.https%3A%2F%2Fwww.gnis.fr%2Fwith+courtesy&rlz=1C1GCEU_enIN1020IN1020&oq=www.https%3A%2F%2Fwww.gnis.fr%2Fwith+courtesy&aqs=chrome.69i57j0i546i649j69i58.233j0j9&sourceid=chrome&ie=UTF-8). (GNIS: Groupement National Interprofessionnel des Semences et plants which changed its name in January 2021 to SEMAE (interprofession of all seeds and for all uses))

---

[4] 1100 million tons in 2019 ahead of wheat 734 million tons and rice 495 million tons (source McCormick).

They contain an average of 400–500 grains (up to 1000 grains) and give yields that can exceed 10 tons per ha. Farmers around the world have continued the varietal improvement work begun by the Amerindian populations, creating thousands of varieties adapted to their climates and eating habits.

Many other crops (wheat, barley, apple, plum, etc.) are also the result of a long varietal selection process adapted to the growing conditions and climate to better satisfy producers and consumers.

These first modifications of the genome, first empirical and then voluntary, are therefore the result of an ancestral human activity. They have made it possible to feed an ever-increasing number of human beings, due to the tremendous human demographic growth that we are experiencing.

Conclusion

The progress of knowledge in genetics, biochemistry, and molecular biology in the nineteenth and twentieth centuries has allowed us not only to understand the evolutionary phenomena of biological organisms but also to develop new techniques so that phenomena produced by chance in nature and which could prove beneficial to humans can be reproduced and selected in the laboratory.

Genetic engineering makes it possible to reproduce in the laboratory phenomena observed in nature, such as mutagenesis and transgenesis, and to modify the genome of organisms to obtain a desired trait. Produced by nature in a random way, the genetic modification that has caught the interest of humans becomes reproducible but can also be improved as scientific and technological advances are made.

Advances in genomic science over the past two decades have shattered the claim that performing horizontal gene transfers in the laboratory would transgress a genetic barrier between species, which is proving to be a myth.

Bibliography

1. Houdebine, L.-M. (2000). *Le vrai du faux* (p. 236). éditions Le Pommier.
2. Le Guyader, H. (2018). Comment le blé est devenu tendre. *Pour la Science, 494*, 92–94.
3. Regnault-Roger, C. (2020). *Des outils de modifications du génome pour la santé humaine et animale* (p. 56). Fondation pour l'innovation politique.

Chapter 4
Evolution of Genome Modification Techniques

Based on the scientific approach, research was developed during the twentieth century to develop more efficient tools for modifying the genome without waiting for a spontaneous chance. Modern genome modification techniques aim to increase the precision of the transformations and to minimize the unintended effects that may accompany these changes or even to anticipate them in order to better prevent them.

Today, the techniques used to modify the genome can be classified into two categories: first-generation techniques developed during the twentieth century and second-generation techniques – new techniques developed after the implementation of laboratory transgenesis and which the European Commission (EC) has just qualified in April 2021 as *New Genomic Techniques* (NGTs), a term that is destined to become more widespread.

Among these, the CRISPR (*clustered regularly interspaced short palindromic repeats*) technique, a revolutionary approach to making changes in genomes, is being praised. Published in the journal *Science* in 2012 [1] by two scientists, Emmanuelle Charpentier, the European, now director of the Max Planck Research Center for Pathogen Science (*Max-Planck- Forschungsstelle für die Wissenschaft der Pathogene*) in Berlin, and Jennifer Doudna, the American professor of biochemistry and molecular biology at the University of California at Berkeley, this major discovery was awarded the Nobel Prize in Chemistry in November 2020.

First-Generation Biotechnologies

Research on microorganisms and plants has proven to be easier to conduct than on higher animals. As mentioned in Chap. 2, the first transgenic transformations were carried out as early as 1973 on bacteria, which are simpler organisms than plants or higher animals because they are single-celled, before being carried out on plants in 1983. Transformations by random mutagenesis are older; they were carried out as

C. Regnault-Roger, *Biotech Challenges*, https://doi.org/10.1007/978-3-031-38237-6_4

early as the 1940s–1950s and were followed by other techniques (e.g., in vitro culture or protoplast fusion) developed in the following decades.

The two techniques very used (transgenesis, mutagenesis) are heavy to implement and not very precise.

Mutagenesis

Random mutagenesis techniques have been developed in the laboratory since the 1940s. Mutations are induced by mutagenic chemical agents (e.g., colchicine, ethyl methanesulfonate) or physical agents (UV radiation, heat). Their goal is to obtain mutants with interesting traits and to select them without waiting for a spontaneous mutation to be observed in the field. However, the technique is random because the mutations are numerous, and only a small number of mutated cells show the desired characteristics. A long, tedious, and costly sorting phase in terms of time and manpower is necessary to select the right mutants. Nevertheless, random mutagenesis makes it possible to obtain improved plant varieties that incorporate the desired trait more quickly than by sexual crossing. Random mutagenesis programs are commonly used by breeders for plant breeding.

These varieties are part of the 9000 registered in the *French Official Catalogue of Cultivated Plant Species and Varieties* and 23,000 agricultural species as well as 21,000 vegetable species and 12,500 fruit species listed in the European Catalogue [2]. All these species and varieties registered in the Catalogue respect quality criteria by passing the DHS/VATE tests.[1] They are used all over the world, including in organic agriculture, without any health risk being mentioned. However, violent campaigns by NGOs have been conducted in France for several years to prohibit the cultivation of varieties obtained by in vitro mutagenesis from a spontaneous mutation observed in the field (see Chap. 5).

Transgenesis

Like random or directed mutagenesis, transgenesis, the result of human activity, makes it possible to select the characteristics that one wants to insert into a given organism and to measure the consequences of this genetic modification. It is controlled and thought out, unlike spontaneous transgenesis. The first transgenic organisms were developed in the laboratory in the 1970s, first on unicellular microorganisms and then on higher multicellular organisms, plants, and animals (see Chap. 2).

Plant transgenesis in the laboratory is a complex process. Three methods are initially used to obtain transformed cells: the particle gun that randomly bombards

---

[1] The DHS/VATE tests characterize varieties by their distinctness, uniformity, and stability (DHS) as well as by their agronomic, technological, and environmental values (VATE). See GEVES https://www.geves.fr/qui-sommes-nous/sev/etudes-dhs-vate/

the cells in which one wants to modify the genome by transgenesis, the electroporation devices that modify the permeability of the cell membranes in order to allow the penetration of genomic material, or the incorporation of the inserted gene fragment of interest in a genetic construction (plasmid) of DNA by the soil bacterium *Agrobacterium tumefaciens* capable of transferring plasmid DNA into plants (cf. Chap. 2). The first two techniques (biolistics and electroporation) have been superseded today by the third, bacterial transfer.

The transformation of the genome takes place in several steps, which makes the whole process somewhat cumbersome. But, in the end, we obtain a plant that has acquired new characteristics, with a considerable saving of time compared to what could be done by the classic techniques. It is not a new species, whatever the origin of the gene that has been transferred, but a new variety. As Philippe Joudrier has pointed out, "A genetically modified maize remains and is always a maize" [3].

Second-Generation Biotechnologies

A lot of research has been done since then to improve the existing genome modification techniques.

New Genomic Techniques (NGT): Definition and Scope

Several techniques were developed in the 1990s, including the use of RNAi (or interfering RNA), which earned its authors Andrew Fire and Craig Mello the Nobel Prize in Physiology and Medicine in 2006.

This RNA, as its name indicates, interferes with the expression of certain proteins and modifies their functional role.

In 2011, the *EU Commission* Joint *Research Centre* (JRC) drew up a non-exhaustive list (Box 4.1), as well as a state of the art of these new emerging techniques [4]. In France, the High Council for Biotechnology (HCB) has subsequently drawn up summary sheets [5].

> **Box 4.1: List of New NGT Techniques Established by the Joint Research Centre (JRC) of the European Commission, Which Are the Subject of HCB Summary Sheets**
> - *Oligonucleotide-directed mutagenesis* (ODM).
> - Meganucleases (MN),
> - *Zinc finger nuclease technology* (ZFN).
> - Grafting, agro-infiltration.
> - *RNA-dependent DNA methylation* (RdDM).
> - *Reverse breeding, synthetic genomics*.
> - *Transcription activator-like effector nuclease* (TALEN).
> - *Clustered regularly interspaced short palindromic repeats* (CRISPR).

Mainly used in plant breeding, these techniques were initially grouped together under the term NPBT (*new plant breeding techniques*) and then under NBT (*new breeding techniques*) because they could be applied to organisms other than plants. This is what the HCB emphasizes, indicating that although they were "grouped together under the same acronym for historical and legal reasons," they are in fact heterogeneous in their mechanisms of action and in their fields of application. It specifies that:

> The application of *site-directed* nucleases (*SDNs*) is also highly developed in the animal world and in medicine.

What Nils Braun, former HCB project manager, summarizes as follows:

> First of all, we should question the very notion of "new plant breeding techniques" (NPBT). Indeed, these "techniques" are, for the most part, not really new, nor do they apply exclusively to plants. Finally, they are not all "techniques." They are, in fact, a "laundry list" not only of techniques, but also of new strategies using already existing techniques [6].

Today, the European Commission (EC) has preferred the more general name of NGTs (*new genomic techniques*), which is indeed a more appropriate name because it emphasizes that they are not only targeted at plant variety selection. This new name is thus a welcome clarification.

However, it would appear that the CJEU ruling of July 25, 2018 (see Introduction), still obscures the scope of the techniques considered under this term. Indeed, in spring 2021, the JRC produced two bibliographic reviews to help the European Commission in its reflection. The first one deals with the state of the art of the new genomic techniques [7] and the second with the commercial applications of these NGTs [8]. Both studies refer to the administrative and legal decision of the CJEU in order to define which techniques fall within the scope of NGTs and which are excluded. It follows that only techniques developed after the date of publication of Directive 2001/18/EC, which still governs the regulatory status of GMOs in the European Union today, 20 years later, are classified as NGTs, based on a definition of GMOs that is now disputed. The JRC distinguishes two categories: "*established genomic techniques*" (EGTs), which are techniques described and developed before 2001, and "new *genomic techniques*" (NGTs), which are techniques developed after 2001.

This is scientific nonsense, since there is a continuum in the discovery of a technique, which is always in its infancy, and the improvements that are made in the years following implementation. Classifying genomic modification techniques according to the date of publication of an administrative regulation and a court decision by the CJEU has left many biotech specialists stunned and is not in keeping with any scientific logic. This is a simplistic legal approach to the detriment of the complexity of the evolution of scientific thought. This approach obviously does not bring the necessary clarity to the debate.

The Meteoric Rise of NGTs

The rapid advancement of knowledge about cellular genomic functionality and the innovations derived from it cannot escape anyone in this post-COVID-19 pandemic period, which has seen the emergence of new technologies to create messenger RNA-based vaccines against COVID. The evolution of genome-editing techniques is part of this movement of constant change and improvement.

Several of these genomic modification techniques, such as the meganucleases discovered in the late 1980s and that of zinc finger nucleases (ZFNs) described in 1996, are already considered to be underperforming and too expensive in light of the latest discoveries. A report by the French Parliamentary Office for the Evaluation of Scientific and Technical Choices (*Office parlementaire pour l'évaluation des choix scientifiques et techniques* OPECST) published in 2017 [9] emphasized the disparity in the costs of implementing these techniques, which it quantified at €50,000 for an intervention using meganucleases, a technique that has now been virtually abandoned, €5000 for ZFNs, €1000 for TALEN, and €10 for the synthesis of RNA that is the guide to the technique that is now favored by all, the CRISPR technique, which has been described as "molecular scissors" to emphasize its precision or "garage biology" because of its ease of implementation and lower cost.

In reality, this technique, while easy to perform and less expensive than others, also requires high-tech environment and experienced personnel. However, it is much more affordable than its predecessors, ZFN and TALEN or ODM, which it quickly supplanted. A great deal of research has been devoted to it between 2012 and April 2021 resulting in over 20,000 scientific publications, which is considerable (see Chap. 13). This observation is confirmed by the French High Council for Biotechnology, which emphasizes that the CRISPR technique is by far the most widely used (68.5% of cases), ahead of TALEN (8.4%) and ZFN (6.9%) [10].

This CRISPR technique associates a guide RNA with a protein (of the Cas type, initially Cas9) and makes it possible to cleave the DNA at a very specific point by a double-stranded cut, causing a modification of the gene with great precision. As a corollary, it is accompanied by a reduction in the number of mutants induced, since they are divided by a factor of 100, according to Georges Pelletier, a member of the French Academy of Sciences and the French Academy of Agriculture, which makes the final sorting phase lighter and saves time and money.

However, new genome-editing techniques derived from CRISPR are already emerging:

- Thus, *base editing*, described in November 2017 by David Liu's team from Harvard University, allows with a CAS-type nuclease associated with another enzyme to transform a DNA base into another, thus inducing a targeted mutation, without having to break the DNA structure. This technique, which is precise and light in the genetic modification it performs, raises great hopes for the repair of point mutations that cause genetic or tumor diseases. Liu likens basic editing to a pencil that erases imperfections.

- *Prime-editing*, also published by the same team more recently in October 2019, is based on an enzyme (a *reverse transcriptase*) and a *"prime-editing"* guide RNA (pegRNA), which allows for the controlled replacement of any base in the genome by insertion, deletion, or base-for-base conversion.

These two improvements of the CRISPR technique also prove to be highly accurate and cause minor changes in the genome, however, with major consequences in terms of improved health for patients with blood diseases such as β-thalassemia or sickle cell disease, as highlighted in January 2021 by Annarita Miccio's INSERM-Imagine Institute (Paris) team [11]. These hemoglobinopathies currently affect more than 130 million humans and are a major public health issue.

Classification of NGTs

According to the way in which the modifications on the nucleic acids that make up the genome take place, the NGTs have been divided into four categories by the JRC [7]:

- Category 1: DNA double-strand breakage by *site-directed* nucleases (*SDN*) with MN, ZFN, TALEN, and CRISPR/Cas techniques. These techniques can lead to mutagenesis, cisgenesis, intragenesis, and targeted transgenesis events. These modifications account for 90.6% of the reported cases.
- Category 2: Single-strand break or without any DNA break – with ODM, *base editing*, and *prime-editing* techniques – which concerns 7.7% of the cases.
- Category 3: No genome sequence modification but gene expression (epigenetic) modification with the RdDM technique for 0.8% of cases.
- Category 4: Action on RNA with the *RNA-editing* technique – less than 0.5% of cases.

However, instead of this classification by type of technique used, a classification based on the modification obtained from the DNA after the action of the SDN enzymes is preferred.

Three classes are thus defined, called SDN1, SDN2, and SDN3:

- SDN1 refers to the inactivation of a gene by deletion or random insertion and substitution of one or more bases.
- SDN2 refers to the replacement of one allele or sequence of an allele by another; there is no stable insertion of recombinant DNA.
- SDN3 refers to a DNA double-strand break repaired with an exogenous double-stranded DNA. The products obtained are considered to be the result of transgenesis with however a much more precise control of the transgene insertion site.

Most products obtained by SDN3 are detectable by the methods used for conventional transgenesis, which is not the case for SDN1 and SDN2. The products obtained by ODM, *base-editing*, and *prime-editing* techniques are considered as SDN1 and SDN2.

Genome modifications of the SDN1 type, and in a large majority SDN2, are very common in nature. The genetic examination of the organisms obtained does not make it possible to know if it is a spontaneous modification in vivo or a modification carried out in a laboratory in vitro. Under these conditions, it is impossible to say that a product obtained by NGT is different from a product obtained by conventional techniques or even discovered in nature, making any biological traceability inoperative. This observation has serious consequences for the regulations that must be applied to them, because if it cannot be distinguished from a product obtained by other techniques, why apply different regulations?

The SDN classification considering the characteristics of the genomic transformation of the final product is now used in many countries to define the regulations to be applied to NGTs.

Conclusion

Modern biotechnology now has a vast array of tools for modifying the genome. Although they have broadened approaches to health, agriculture, and the environment, as new vaccines against COVID have vividly demonstrated, they have not developed without opposition despite the scientific advances they have enabled.

Is there any truth to this societal reluctance?

Bibliography

1. Jinek, M., Chylinski, K., Fonfara, I., Hauer, M., Doudna, J. A., & Charpentier, E. (2012). A programmable dual-RNA-guided DNA endonuclease in adaptive bacterial immunity. *Science, 337*(6096), 816–821. https://doi.org/10.1126/science.1225829
2. SEMAE. (2021). *Les catalogues officiels français et européen des variétés*, https://www.semae.fr/catalogue-varietes/
3. Joudrier, P. (2012). Création et amélioration variétale- transgenèse et sécurité alimentaire. In H. Regnault, X. Arnauld de Sartre, C. Regnault-Roger (coord), *Les Révolutions agricoles en perspective* (pp. 119–138). Editions France Agricole.
4. Lusser, M., Parisi, C., Rodriguez Cerezo, E., & Plan, D. (2011). *New plant breeding techniques. State-of-the-art and prospects for commercial development* EUR 24760 EN. Luxembourg: Publications Office of the European Union; 2011. JRC63971, https://publications.jrc.ec.europa.eu/repository/handle/JRC63971
5. High Council for Biotechnology. (2016). *Nouvelles techniques Première étape de la réflexion du HCB. Introduction générale*, Rapport provisoire sur les NPBT du CS, 107 pages, http://www.hautconseildesbiotechnologies.fr/fr/article/publications-hcb
6. Braun, N. (2017). Portée et limites des nouvelles techniques d'obtention végétale, les New Plant Breeding Techniques (NPBT). *Annales des Mines-Réalités industrielles, 2017*(1), 90–93.
7. Broothaerts, W., Jacchia, S., Angers, A., Petrillo, M., Querci, M., Savini, C., Van den Eede, G., & Emons, H. (2021). *New genomic techniques: State-of-the-art review*. EUR 30430 EN, Publications Office of the European Union, Luxembourg. https://doi.org/10.2760/71005 6JRC121847

8. Parisi, C., & Rodríguez-Cerezo, E. (2021). *Current and future market applications of new genomic techniques.* EUR 30589 EN, Publications Office of the European Union, Luxembourg. https://doi.org/10.2760/02472JRC123830

9. Le Déaut, J. -Y., & Procaccia, C. (2017). *Les enjeux économiques, environnementaux, sanitaires et éthiques des biotechnologies à la lumière des nouvelles pistes de recherche* OPECST information report, N°507 April 2017, https://www.senat.fr/notice-rapport/2016/r16-507-1-notice.html

10. High Council for Biotechnology. (2021, November 26). *Synthesis on the detection of products from new genomic technologies (NGT) applied to plants,* Report of the Scientific Committee, http://www.hautconseildesbiotechnologies.fr/fr/article/publications-hcb

11. Imagine Institute. (2021). *Team Annarita Miccio,* https://www.institutimagine.org/fr/annarita-miccio-189

Chapter 5
At the Heart of a Societal Controversy

Unaware that they have been surrounded by and consuming products of biotechnology for centuries, the general public became wary of them in the late 1990s, particularly those resulting from genetic engineering. In France, this mistrust was triggered by militant NGO campaigns, suspicious or malicious media allegations, and uninformed choices by several political leaders in positions of authority who, out of conviction or lack of courage, decided to sacrifice biotechnological innovation. Green biotechnologies have been particularly targeted.

Since then, agricultural biotechnologies in France and Europe have been in a state of perdition. It is a descent into hell that has lasted for several decades, and several events testify to this.

When "Crazy Soy" Meets Greenpeace (1996)

The key event in France that sparked off the controversy was the front page of the newspaper *Libération*, which ran the headline *Alerte au soja fou* on November 1, 1996. This headline (Crazy/Mad Soy Alert !) referred to the "mad cow crisis" that Europe was experiencing at the time and which was, according to Marcel Kuntz, "the defining crisis for the fate of GM plants in Europe" [1]. Indeed, in the 1980s and 1990s, *the contaminated blood affair* and *the mad cow disease crisis* undermined public confidence in the public authorities responsible for ensuring their health safety. This "mad soy" is Monsanto's transgenic soy from the United States, which received, along with Novartis' transgenic corn, an authorization for import from the European Union in February 1996.

Jean-Claude Jaillette, responsible for the headlines of the newspaper *Libération*, agreed that it was a question of reporting on the importation of American transgenic soybeans into the ports of Rotterdam and Bordeaux [2], thereby supporting the

C. Regnault-Roger, *Biotech Challenges*,
https://doi.org/10.1007/978-3-031-38237-6_5

campaign that the NGO Greenpeace had just launched on the subject. Noting that its finances were in bad shape after the end of the French nuclear tests in Mururoa and that its positions against the Gulf War had caused more than a million American donors to flee, the NGO was in fact wondering what battle horse could replenish its coffers[1]. An informal group gathered in Europe around Isabelle Meister and Arnaud Apotecker concluded "that the fight against GMOs would be a promising theme" [3] and they decided to launch a campaign against the import of American soybeans into Europe[2].

However, the French management of the NGO is not convinced that GMOs are a real danger to health, since as Bernard le Buanec, member of the Academy of Agriculture and the Academy of Technologies, reports [3], Bruno Rebelle, director of Greenpeace France in 2002, declared:

> We are not afraid of GMOs. We are only convinced that it is a bad solution. GMOs may be a wonderful solution for a certain type of society. But it is precisely the kind of society we do not want.

Fifteen years later, anti-biotech NGOs still hold this view. The international association of the political ecology movement "Friends of the Earth" states in 2016 about the new genomic techniques:

> The choice that will be made on new techniques will decide for or against a societal model that seeks in a technological headlong rush the solution for the ills of the unbridled consumer society and climate change.

It specifies that it is a refusal of innovation and invokes the precautionary principle:

> The decision that must be made about new techniques is a political choice, not a technical one. Do we maintain the precautionary principle, or do we embark on a blind and frantic race that seeks salvation in novelty and in an unconditional belief in progress," and she adds, "What is at stake is not this or that technique but the society that develops them". [5, 6]

In the autumn of 2021, the discourse of a number of prominent political ecologists, including two candidates for the EELV (*Europe Ecologie Les Verts)* presidential nomination in 2022 (Delphine Batho and Sandrine Rousseau), has become more decisive and claimed to be about economic degrowth.

The rejection of genetic engineering, GMOs, and genome-editing techniques professed by some NGOs and other environmental lobbyists is part of this ideological movement and is not based on scientific arguments. To ask them to examine the results of the last 20 years in order to convince them that the benefits outweigh the risks is a perfectly illusory approach.

---

[1] It is interesting to read La Crise n°50, a socioeconomic chronicle posted on the website of the Université du Québec à Montréal (UQAM), which examines the links between the actions of Greenpeace and its financial operations (page 9) (https://www.ieim.uqam.ca/spip. php?page = article-ceim&id_article = 13175).

[2] Hervé Kempf's book tells the story behind the scenes of these decisions [4].

These NGOs, which are fiercely opposed to the development of agricultural biotechnologies, clearly display an ideological opposition to the current society, and this orientation occurred very early. It is therefore political points of view that are put forward on a model of society and not scientific considerations.

Tribulations of GMO Corn Crops in France (2007–2014)

Among the twists and turns that have accompanied the deterioration of the situation of biotech plants in France, the one that led to the banning of transgenic corn in France in 2014 and the bending of European regulations on GMOs in 2015 deserves to be recounted.

Indeed, until this date (2015), when a European Community authorization was granted for the import or cultivation of a genetically modified plant, in accordance with European regulations (Directive 2001/18/EC) to which France is subject due to its membership of the European Union (EU), the EU Member States could only oppose this decision by invoking a "safeguard clause," which was later entitled "emergency measure." This is based on scientific evidence of the existence of a threat to public health or the environment, related to the specific situation in the country.

This context generated an unprecedented situation in France, not without its share of comicalities, if it had not been for a political deception at the highest level of the State apparatus, which contributed to discrediting scientific expertise and rendering the scientific word inaudible.

In France: The "GMO Versus Nuclear" Deal and Far-Fetched Moratoria

We know today from François Fillon, then Prime Minister of France, that an agreement was reached between President Nicolas Sarkozy and the elected representatives of the environmental movement during the *Grenelle de l'environnement* in 2007:

> He had arbitrated in favour of a total ban on GMOs and open field experiments... their abandonment made it possible to obtain a deal with the ecologists who, if they were granted this, would not obstruct the nuclear issue [7].

But at that time, according to the European regulations in force, there was nothing to prohibit the cultivation and import into France of GMPs authorized in the EU, except to invoke the safeguard clause and decide on a moratorium, the scientific basis of which must be evaluated by the *European Food and Security Agency* (EFSA), which sends an opinion to the European Commission, which then decides.

At that time, a transgenic maize called MON 810 was being regularly cultivated by farmers in the South of France who wanted to protect themselves from the ravages of the European corn borer and the pink stem borer, two lepidopteran borers that cause significant damage to ears and stalks when they are present (see Chap. 11).

From 2008 onward, several ministerial decisions were taken to prohibit the cultivation of maize by invoking moratoria based on reports by experts who spoke outside their field of competence: for example, a specialist in penguins in Antarctica (where, as everyone knows, corn fields stretch as far as the eye can see!) wrote the *Le Maho report* (known by this name, which is that of its author), which was used by the French government to initiate the first moratorium on the cultivation of maize (MON 810) in 2009 [8].

This was followed by a series of other reports that I have described as "scientific nonsense" [9] because the scientific arguments used to justify the moratoria were considered weak. EFSA rejected them four times in 2009, 2011, 2012, and 2014. These rejections, which were scientifically supported, were followed by the Court of Justice of the European Union (CJEU), which annulled the moratoria imposed by successive French governments, a decision that was ratified by the French *Conseil d'Etat* (Council of State), following the CJEU, in a legal logic.

Political Choices Far from Science

These repeated political and legal twists and turns became a farce: the arguments deployed in the various documents produced by the French authorities are worthy of the best plays of Molière and so inconsistent that they become the laughingstock of the expert commissions! But it is a tragic farce because, at the same time, acts of vandalism by a group of anti-GMO activists, the *"Faucheurs Volontaires"* (Volontary Reapers), are ruining the work of the farmer who has legally planted transgenic corn in his fields. Indeed, echoing the European Food Safety Authority (EFSA) the *French High Council for Biotechnology* (HCB) regularly concludes, year after year, that "the analyses contained in Monsanto's monitoring report do not reveal any major problem associated with the cultivation of MON 810 corn." But under the eye of television cameras summoned to immortalize the heroic rampage, and under the benevolent gaze of the public forces who remained passively at the edge of the field, the ecologist Member of European Parliament José Bové and Greenpeace activists vandalized plots of MON 810 corn seedlings that have been sown in full legality.

However, after 6 years of this inextricable situation, it was necessary to get out of the regulatory imbroglio. Indeed, if the presidents of the Republic change, the will to ban the cultivation of GMOs still remains. The first socialist government of Jean-Marc Ayrault (2012–2014), under the presidency of François Hollande, published an order in the *Official Journal of the French Republic* on March 14, 2014, prohibiting the cultivation of the 232 varieties of Bt MON 810 corn seeds registered in the *Official Catalogue of Species and Varieties of Cultivated Plants*. The date was chosen to make illegal the sowing of MON 810 corn planted in the spring of 2014. Farmers in Haute-Garonne, Tarn-et-Garonne, and Gers, having planted MON 810 varieties before the date of the order, were forced by decision of the Minister of Agriculture (then Stéphane Le Foll) to uproot their fields, the Council of State, seized urgently in summary proceedings, having refused to suspend the order of March 14, 2014.

And to further strengthen this position, the second socialist government (Prime Minister Manuel Valls from 2014 to 2016) had a law passed in the wake of this, in an accelerated procedure by Parliament, on June 2, 2014. The Law No. 2014–567, also known as the *small law* because it contains only one article, prohibited the cultivation of genetically modified corn varieties, after the Constitutional Council had ruled it to be in compliance with the Constitution.

Here is the full text:
Single Article:

I. The cultivation of genetically modified corn varieties is prohibited.
II. Compliance with the prohibition on cultivation provided for in I is monitored by the agents mentioned in Article L. 250-2 of the Rural and Maritime Fishing Code. These agents have the powers provided for in articles L. 250-5 and L. 250-6 of the same code. In case of non-compliance with this prohibition, the administrative authority may order the destruction of the crops concerned. This law shall be executed as a law of the State.

This *little law* flouts the seriousness of parliamentary work on two counts: it is based on unfounded scientific arguments, and it is in complete contradiction with European regulations, which its promoters could not have been unaware of [9].

Law 2014–567: An Unfounded Scientific Argument

The *small law* is presented as a bill by Senator Alain Fauconnier (PS). However, the statement of its reasons had challenged Senator Jean Bizet (LR) and Deputy Bernard Accoyer (LR) who then seized, on April 24, 2014, the High Council of Biotechnology (HCB), under the law 2008–595 to collect its scientific opinion, asking four questions. As the HCB had completed its first mandate on April 30, 2014, it was necessary to wait for its renewal before it could respond to this referral, which was done by an opinion of the HCB Scientific Committee published on June 23, 2015. And although the law was voted in the meantime on June 2, 2014, the responses of the HCB Scientific Committee to the parliamentary referral do not fail to question.

The first question asks whether there are indeed environmental problems linked to the cultivation of Bt MON 810 maize (mentioned in a recital of the law) in the countries of the European Union that have been cultivating it for several years, in terms of resistance in the target insects (the two lepidopterans, the European corn borer and the pink stem borer, already mentioned) and in terms of the reduction of populations of nontarget lepidopteran insects that would be sensitive to it. The HCB relies on the opinions published by EFSA in 2009 and 2012, which show that no development of resistance to the Cry1Ab toxin in populations of exposed target lepidopterans in EU countries has been recorded since these Bt maize varieties were cultivated in Europe, and this, with a hindsight of more than 10 years. Therefore, the HCB indicates that no adverse environmental effects of Bt maize MON 810, under the intended uses and recommended management measures in EU countries, have occurred.

The second question concerns the unintended effects that the cultivation of MON 810 maize could have on other insect populations. The HCB emphasizes the difficulties of conducting exhaustive monitoring on all natural populations of species that would be exposed to the maize crop and the methodological limitations of such monitoring. It recalls that the spectrum of species sensitive to the Bt toxin is narrow and that any change in agricultural practices will necessarily have an impact on the species that depend on an agroecosystem but that these changes are not necessarily long term. Consequently, the HCB concludes that the answer to the question is negative, i.e.. "no unintended effect of the cultivation of MON 810 maize on populations of susceptible target and nontarget lepidopterans has been observed in Europe."

The third question is based on the existence of a dominant resistance mechanism to the Cry1Ab toxin observed in a sub-Saharan insect pest, *Busseola fusca*, and asks whether it is likely to cause environmental damage in France. The HCB also answers negatively to this question by noting that this insect is not present in Europe and that "a resistance mechanism cannot be extrapolated from one species to another."

The fourth and last question asks for clarification of the terms of the proposed law, namely, that growing transgenic maize such as MON 810 or Bt 1507 maize poses a "significant risk that clearly endangers the environment." To this last question, the HCB Scientific Committee responds that it did not identify such a risk in its opinions issued in 2009, 2010, 2011, and 2013 and that the rapporteur of the proposed law, Senator Fauconnier, who expressed this particular view, should be approached to support it.

Thus, this proposed law that was voted on June 2, 2014, by the French Parliament has no scientific basis nor does the scientific argument of the March 14, 2014, order banning the marketing, use, and cultivation of the 232 seed varieties of Bt MON 810 maize, an argument that was judged severely by EFSA in August 2014.

It must be concluded that the environmental risks of the explanatory memorandum of law n°2014–567 of June 2, 2014, are not proven, whereas, on the contrary, numerous studies have shown that the cultivation of MON 810 has, for example, a beneficial effect in regions where *Fusarium* wilt occurs (reduction of mycotoxin levels and the incidence of the diseases they cause). In southern European countries, the mycotoxin risk is an emerging risk favored by global warming (see Chap. 11).

An Epilogue: Modification of the European Regulation by the Directive (EU) 2015/412

While it was pushing through the "little law" in France, the French government was actively lobbying for the adoption of a new regulation in Brussels that would allow it to justify its action *after the fact*. The EU Council reached a political agreement on June 12, 2014, discussed by the environment ministers of the various countries. Its aim is to give Member States the possibility of restricting or banning the cultivation of genetically modified organisms (GMOs) on their territory, citing environmental but also socioeconomic reasons. From now on, the adopted Directive 2015/412 authorizes a Member State to take:

> "measures restricting or prohibiting, in all or part of its territory, the cultivation of a GMO or a group of GMOs" provided that "such measures are in accordance with Union law, are reasoned, proportionate and non-discriminatory and, furthermore, are based on serious grounds such as those relating to: environmental policy objectives; land use planning, land use; socio-economic impacts; the desire to avoid the presence of GMOs in other products" (with appropriate measures for border areas Editor's note); agricultural policy objectives; public order, this last point to be raised in conjunction with another reason.

On the strength of these new provisions, the French government (that of Manuel Valls but also all the following ones) hastened to assert that it was opposed to any cultivation of genetically modified plants in France and this, henceforth, in full European legality! This put an end to the story of the cultivation of Bt MON 810 corn, which was first authorized and then prohibited in France.

As I have pointed out [9]:

> We will remember that the cultivation of genetically modified plants is prohibited in France for political alliance considerations, and that the European Commission has accepted that a regulation is no longer based solely on scientific criteria but on the political and ideological positions of the governments of each Member State, thus leaving the field open to subjectivity and arbitrariness.

A Long List of European Renunciations

As everyone knows, the cultivation of biotech plants has become very marginal in Europe with the notable exception of Spain and Portugal, where the only existing GMO crop is MON 810 transgenic maize, practiced on 111,883 ha in 2019 [10], however, a steady decrease since the year 2013, when the areas reached 136,962 ha (and 120,990 ha in 2018). As of June 2019, the MON 810 transformation is present in 150 varieties out of the 5479 maize varieties listed in the *European Official Catalogue of Cultivated Plant Varieties*. The regions where it is grown (Aragon, Catalonia, Extremadura, Spain, and Alentejo and the Lisbon region in Portugal) are subject to strong pest pressure. Thus, there has been a small decline in GMO crops in Spain and Portugal, which may be due to a temporary or cyclical situation, as the future will show (see Chap. 16).

In other EU countries, there has been a long series of renunciations of the cultivation of this corn but also of other transgenic plants. In fact, in the 2010s, MON 810 maize was also grown in the Czech Republic and Slovakia for livestock feed and biofuel. Among the abandoned crops was the cultivation of the transgenic potato *Amflora* in Germany, Sweden, and the Czech Republic: its high amylopectin starch, which has interesting properties for industrial compounds (packaging, adhesives, textiles, etc.) and starch products, optimized the treatment processes by saving energy and water. Ransacking of the plots in 2011 and 2012 by activists led to the abandonment of this crop in Europe. In 2012, the company BASF, which marketed this transgenic potato, decided to relocate this activity to the United States.

Since then, several French and foreign companies have followed this example and transferred their research and development activities in this field to the American continent, North and South. And when a varietal innovation obtained thanks to genome modification techniques comes to market, international seed companies no longer even consider European needs and turn to the Asian and American markets that are favorable to biotech products.

An illustration of this delocalization is provided by the world's first transgenic wheat, HB4, which received marketing authorization in Argentina in early 2023. The company Trigall, which is responsible for this product, is a joint venture between the Argentine company Bioceres and the French seed company Florimond-Desprez. The latter has been developing research activities for several years in this South American country, which is more welcoming to biotechnological innovations.

Today, most EU countries reject the cultivation of biotech/GMO plants after having accepted it. In addition to various reasons linked to the different national contexts, there is above all the relentless action of anti-GMO movements, such as Greenpeace and Friends of the Earth, and in France the actions of the Faucheurs volontaires (Voluntary Reapers) and the agricultural union named La Confédération Paysanne.

Bernard Le Buanec in his book *Les OGM Pourquoi la France n'en cultive plus?* (GMOs Why doesn't France grow them any more?) [3] analyzed what led to this situation of refusal of GMOs. He notes a convergence of attitudes of various social actors:

- Regular destruction of experimental trials and ransacking of legal cultivated plots without court sentences.
- The marketing choice of the Carrefour Group, soon to be followed by several other supermarket chains, to label its products with the "GMO-free" label as if GMOs were something dangerous.
- The procrastination of INRA (now INRAE) in the years 1990–2000, which did not adopt a clearly favorable position.
- Agricultural unions that were divided or hesitant at the time (*FNSEA*), with the *Confédération Paysanne* radically opposed to GMOs, and also at this moment the *Coordination Rurale*.

- The political *deal* of 2007 at the highest level of the French state mentioned above.
- The recurrent media campaigns, the one of *Libération* mentioned above but also a surprising manipulation of the newspaper *Le Nouvel Observateur* of September 19, 2012, titling *"Oui les OGM sont des poisons"* (Yves, GMOs are poison). This assertion was based on a publication that was subsequently judged to be contentious and has since been retracted (i.e., withdrawn from publication by the management of the international journal *Food and Chemical Toxicology*) in the face of the avalanche of negative opinions from all national and international agencies. Although it has been invalidated by research carried out since then by impartial public bodies (for the modest sum of 15 million euros!), this study has done considerable damage to public opinion in France and the rest of the world [11]. It is therefore not surprising that in a 2018 IPSOS survey, 75% of French people said they were concerned that their food should not contain GMOs [12].

The "Hidden GMOs"

Today, due to the lack of transgenic plant fields to destroy since the ban on their cultivation in France in 2014, the actions of the Voluntary Reapers target crops derived from mutagenesis, which they call "hidden GMOs."

The European legislator reserved the EU regulation (see Chap. 7) only for products obtained by transgenesis, the latest technique discovered in the 1980s and 1990s, while those obtained by random mutagenesis, a technique used since the 1940s, were exempted, because, despite the precautionary principle on which Directive 2001/18/EC is based, it might have seemed superfluous, even then, to check the results of a technique that had never given rise to any problems for more than 50 years.

Opponents of GMOs had contested this provision and had taken action through the courts and through illegal and violent actions to destroy crops in the field. As Jean-Yves Le Déaut [13] points out, the opponents of GMOs have a well-tried method that complements the ransacking of cultivated plots: to make the regulations more complex in order to cause regulatory stalemate.

These abuses, particularly in France, benefit from a great judicial leniency. Currently, there are no more experimental GMO field trials, whereas 20 years ago there were nearly 800.

The issue of "hidden GMOs" has recently been legally settled. The European Court of Justice (CJEU) issued a ruling on February 7, 2023, in which it recalled that varieties obtained by random mutagenesis techniques in vivo as well as in vitro must be excluded from the regulations governing the cultivation of genetically modified plants in the EU. It rejects the erroneous interpretation made by the French *Conseil d'État*.

Organized Vandalism

Today, the "Faucheurs volontaires" are attacking fields where plant varieties regularly registered in the *French Catalogue of Species and Varieties* are growing and destroying equally legal seeds stored in the closed premises of cooperatives. Among the recent ransacking are noted the destruction of agroecological experimental trials carried out by technical institutes (*Arvalis-Institut du Végétal, Institut Technique de la Betterave*, and also *Terres Inovia*) on an experimental platform in Montesquieu-Lauragais on August 17, 2021 [14]; bags of seeds stored in the premises of *Top Semences*; a union of 11 cooperatives located in the Southeast of France, on June 14, 2021; or the destruction of legally cultivated fields in Ambeyrac in Aveyron on August 22, 2021, with sunflowers of the variety *Vollcano CLP* of RAGT[3], a variety with multiple properties: high oleic acid content, an unsaturated fatty acid with an interesting nutritional profile, mildew resistance, and herbicide tolerance to facilitate weed control and fight against ragweed, a highly allergenic weed.

Intimidation in laboratories is not the only one. Research and development laboratories of seed companies (*Limagrain*, near Clermont-Ferrand) are also the target of these anti-GMO movements who occupy the research laboratories to disrupt the work in progress. Public academic research is also the target of these contemptuous people: the premises of the INRA and those of the *École Normale Supérieure* (ENS) of Lyon (laboratory Reproduction and development of plants) have been invaded by this same movement. These *Science Reapers*[4] want to spread fear to discourage researchers and dissuade students in plant biotechnology from pursuing their training.

Consequences

The consequence of these actions is a relocation of activities outside the European Union. International companies such as BASF,[5] recognizing the impossibility of conducting their field trials in the EU, have moved across the Atlantic. Others cite reasons related to EU regulations, and still others cite the lack of legal sanctions when activist raids are conducted in the field, in laboratories, warehouses, and companies. By going to the American continent but also, since the Brexit, to Great Britain, which opens its arms wide to them, the actors of the biotechnology sector are fleeing insecurity and European blockades.

---

[3] RAGT (*Rouergue Auvergne Gévaudan Tarnais*) is a hundred-year-old seed group, founded by farmers from Aveyron.

[4] The expression Faucheurs de Science ("*Reapers of science*") is the title of a book by Gil Rivière-Weckstein [15] which details the sometimes devious actions of these activists.

[5] BASF, whose abbreviation originally stood for *Badische Anilin- & Soda-Fabrik*, is a German chemical group considered in 2021 as the largest chemical group in the world.

As a corollary, the brain drain noted by the OPECST report of June 2021 obliterates the dynamism of the sector. The careers of brilliant scientists such as Emmanuelle Charpentier and Catherine Feuillet outside of France; the fate of cutting-edge biotech start-ups such as *Calyxt* (originally *Cellectis Plant Sciences*), which was forced to relocate to the United States; and the poignant testimony of Luc Mathis, President of *Meiogenix*, at the OPECST public hearing of March 18, 2021, are all examples of the deleterious consequences of the lack of support for the entrepreneurial development of biotechnology in France.

Conclusion

This long list of destruction of farmers' work, intimidation of scientists, judicial laxity, resignation, or political maneuvers has created a deleterious situation in France and in the European Union for the development of biotechnologies.

During the 12 years (2009–2021) that I have been a member of the High Council for Biotechnology, I have noticed that the number of applications for the cultivation of genetically modified plants in the EU has fallen steadily and that only applications for authorization to import transgenic seeds for food and feed processing have been examined in recent years.

Under these conditions, how can we continue to refer to a French or European agri-food independence that is shrinking like a stone? We can only agree with the words of Jean-Claude Pernollet, member of the French Academy of Agriculture, who emphasized that "the perennial blockage of the debate on GM plants is particularly harmful to the agricultural future of countries that refuse to cultivate them" [16]. Will the regulatory fate of NGTs in the EU be able to counteract this pessimistic observation?

Bibliography

1. Kuntz, M. (2014). *OGM une question politique, GMO a political question* (p. 143). Presses Universitaires de Grenoble.
2. Jaillette, J. C. (2009). *Sauver les OGM* (p. 246). Hachette.
3. Le Buanec, B. (2016). *Les OGM. Pourquoi la France n'en cultive plus? GMOs. Why doesn't France grow them anymore?* (79). Presses des Mines.
4. Kempf, H. (2003). *La guerre secrète des OGM* (p. 298). Le Seuil.
5. Les Amis de la Terre/ Friends of the Earth. (2016). *HCB, les Amis de la Terre prennent leurs responsabilités*, http://www.amisdelaterre.org/IMG/pdf/retour_sur_l_activite_du_hcb_par_les_at.pdf (online 24 Aug 19).
6. Regnault-Roger, C. (2020). *GMOs and genome-editing products, regulatory and geopolitical issues* (p. 56). Foundation for Political Innovation.
7. Fillon, F. (2016). *Faire* (p. 320). Fayard.

8. Comité de Préfiguration d'une Haute Autorité sur les Organismes Génétiquement Modifiés (2008). *Réponse à « l'analyse réalisée le 30 janvier 2008 par la Société Monsanto de l'avis sur la dissémination du MON 810 sur le territoire français rendu par le Comité de Préfiguration d'une Haute Autorité sur les Organismes Génétiquement Modifiés*, document written by Professor Yvon Le Maho, scientific contact designated to EFSA, www.vie-publique.fr (February 13, 2009).

9. Regnault-Roger, C. (2018). La réglementation au cœur des débats. In C. Regnault-Roger, L. M. Houdebine, & A. Ricroch (dir), *Au-delà des OGM* (pp. 135–164). Presses des Mines.

10. ISAAA. (2021). Global Status of Commercialized Biotech/GM Crops in 2019, *ISAAA Brief* n° 55, https://www.isaaa.org/

11. Kuntz, M. (2019). *The Séralini affair. The impasse of activist science* (p. 59). Foundation for Political Innovation Study.

12. Opinion Valley Survey for IPSOS. (2018). *Le regard des Français sur l'agriculture*, https://www.ipsos.com/sites/default/files/ct/news/documents/2018-11/ipsos_pour_opinion_valley_-_agriculture.pdf

13. Le Déaut, J.-Y. (2021). Innovation et agriculture: les techniques d'hier ne résoudront pas les problèmes de demain. *Revue Paysans et Société, 388*(4), 6–13.

14. SYPPRE. (2021). *Communiqué de presse Plateforme Syppre vandalisée à Montesquieu-Lauragais (31) ARVALIS – Institut du végétal, l'Institut Technique de la Betterave et Terra Inovia portent plainte Press release Syppre platform vandalized in Montesquieu-Lauragais.* https://www.arvalisinstitutduvegetal.fr

15. Pernollet, J. C. (2018). Les plantes génétiquement modifiées. In C. Regnault-Roger (dir), *Idées reçues et agriculture. Parole à la science* (pp. 187–205). Presses des Mines.

16. Rivière-Weckstein, G. (2012). *Les faucheurs de science* (p. 128). Le Publieur.

Chapter 6
GMOs: A Regulatory Concept

For many people, GMOs have become synonymous with monstrosity, whereas, as we have seen in the previous chapters, research after the creation of genetically modified organisms in laboratories by genetic engineering techniques and the establishment of specific regulations in many countries have demonstrated that organisms resulting from mutagenesis or transgenesis exist spontaneously in nature. So why have we created this concept of GMO and what does it cover?

Asilomar

As soon as the first gene transfers were successful (see Chap. 2), Paul Berg, a biochemist at Stanford University, and his American colleagues wanted to bring the scientific community together to compare experiences and discuss these new developments.

At his instigation, 150 specialists from the countries most advanced in the use of molecular techniques of genome modification (about 100 Americans and 50 Europeans and Soviets[1]) met in California at Asilomar in 1975.

The objective of this conference was to discuss whether manipulations with recombinant DNA present risks of accidental release and whether safety measures should be observed to limit possible unintended effects. The discussion focused on the characteristics of the techniques used and was limited to precautions to be observed in the laboratories when the question of legal liability related to the commercialization of these new genetically modified products and possible damage claims in the American courts was raised [1].

---

[1] Era of the USSR (Union of Soviet Socialist Republics).

© The Author(s), under exclusive license to Springer Nature Switzerland AG 2023
C. Regnault-Roger, *Biotech Challenges*,
https://doi.org/10.1007/978-3-031-38237-6_6

As a result of this conference, specific measures were taken by the public authorities, and new rules were enacted to define which manipulations were concerned and what measures should be observed. Regulations on GMOs were put in place in various countries. In the United States, the *Coordinated Framework for Regulation of Biotechnology* was published in 1986. This was followed at the European level by Directives 89/219/EEC and 90/220/EEC on the use of GMOs in contained and open environments, published in 1989 and 1990, and 10 years later, in 2001, by Directive 2001/18/EC, which still governs the use of GMOs in the European Union.

However, in France, as early as 1986, the *Commission de Génie Biomoléculaire* (CGB), chaired by Axel Kahn and then by Marc Fellous, was charged with analyzing the potential health and environmental risks associated with open GMO experimentation.

GMOs: A Regulatory Concept That Differs from Country to Country

To legislate on these new genetically engineered products and set rules, it was necessary to define what we were talking about.

In each country, the regulatory texts therefore gave a definition of what was going to be regulated and consequently of what was to be considered as a GMO.

We have chosen two examples, the European Union and Canada, which have a diametrically opposed conception of the management of GMOs, the first based on the notion of risk, therefore presuming harmful effects, and the second on the notion of benefits in relation to the existing.

To do this, we compared the regulatory texts, both of which are published very officially in French: the European Union publishes its regulatory texts in all the languages of the Member States, and Canada is a constitutionally bilingual English-French country.

Definition of a GMO for the European Union

Classifying GMOs according to the technique used to obtain them (transgenesis, mutagenesis, in vitro fertilization, diploid induction, etc.), Directive 2001/18/EC defines a GMO as:

> an organism, with the exception of human beings, in which the genetic material has been altered in a way that does not occur naturally by mating and/or natural recombination (Article 2),

stating in its preliminary consideration 17 that:

> This Directive should not apply to organisms obtained through certain techniques of genetic modification which have conventionally been used in a number of applications and have a long safety record.

The Directive has thus determined its regulatory scope by providing a definition of what is a genetically modified organism (GMO) and which GMOs will be subject to it and which will be exempt. Only products containing recombinant DNA by transgenesis are ultimately subject to the regulations set out in Directive 2001/18/EC. Organisms produced by the use and marketing of organisms transformed by genetic engineering techniques that are "traditionally used" and "proven safe" are exempted. When there is no identified risk, since safety is proven, the use and marketing of organisms transformed by these techniques of genetic engineering will be subject to the ordinary regulation of marketing authorizations and not to the specific regulation that concerns only those GMOs for which it is suspected that there could be risks for health and the environment (cf. Chap. 7). As the first transgenesis was only carried out in laboratories from 1972 onward, the legislator judged that a 20-year delay for Directive 90/220/EEC, and 30 years for Directive 2001/18/EC, was not sufficient to ensure the safety of this technique!

This is why in the European Union (EU) applications for import or cultivation authorizations only concern products genetically modified by transgenesis, other organisms obtained by mutagenesis or other techniques such as protoplast fusion[2] or diploid induction[3] being exempted if they do not originally contain recombinant DNA.

Thus, the term GMO, which has become part of everyday vocabulary, refers only to products of transgenesis. This is what the OPECST report [2] published in June 2021 under the direction of Senator Catherine Procaccia (Les Républicains) and Deputy Loïc Prud'homme (La France Insoumise) notes:

> The term GMO, which could be understood as any organism whose genome has been modified by the hand of man, refers in fact only to the product of transgenesis, which consists in inserting a gene, possibly from another species, into the genome of a plant.

GMOs in the European Community are thus defined by the method of production and not by the properties and characteristics of the product obtained.

Definition of a GMO in Canada

A genetically modified organism (GMO) is defined by *the Commission de l'éthique en science et en technologie du Québec* as:

---

[2] Protoplast fusion or somatic fusion allows the fusion, from plant cells deprived of their outer cell wall, of the cell content of two different species to form a new hybrid inheriting genetic properties from both original species.

[3] Diploid induction causes the doubling of cell chromosomes under the effect of a chemical agent such as a 0.5% colchicine solution.

a microorganism, plant or animal that have been genetically engineered to possess characteristics that it does not possess at all or that it possesses to a degree that is considered
unsatisfactory in its natural state, or to remove or attenuate characteristics that are considered undesirable.

This definition is completed by the service of the Ministry of Agriculture, Fisheries
and Food of Quebec (MAPAQ) in charge of information on GMOs, which provides
the following precision:

Thus, a GMO is a living being whose genetic material has undergone a specific transformation by a method called transgenesis. To date, the GMOs approved in Canada are plants or
micro-organisms. For example, Bt corn, a GMO for animal feed, is resistant to an insect
pest (the European corn borer), while a GM bacterium secretes a human insulin used to treat
diabetes [3].

This Canadian definition, as we can see, proceeds from a state of mind totally
opposed to the European definition. Also focused on transgenesis, it emphasizes the
positive character of the transformations carried out and the expected benefits without focusing on supposedly unknown, unintended, and unproven risks (see Chap.
7). This is why, according to the *Canadian* Food *Inspection Agency* (CFIA):

In Canada, GMOs are regulated in the same way as conventionally produced agricultural
products [3, 4].

New plant varieties are not evaluated based on their breeding technique but on the
characteristics of the new product submitted for marketing. These are evaluated on
a case-by-case basis according to the regulation "Plants *with novel traits*."

One can conclude from these comparisons that the denomination of GMO is
consequently a regulatory concept which is not universal since it can differ from one
country to another according to legal definitions which are national.

Regulatory Concept and Scientific Definition at Odds

André Gallais, emeritus member of the French Academy of Agriculture and honorary professor at the *Institut National Agronomique de Paris* (now *AgroParisTech*),
has pointed out that the regulatory definition of a GMO is far removed from the
biological definition [5].

He wants to prove that the soft wheat of the Renan variety is indeed a genetically
modified organism in a nonnatural way with crosses between different species that
do not occur spontaneously in nature and a chromosomal doubling obtained with a
chemical agent (Box 6.1). As such, Renan wheat is indeed an artificially obtained
wheat through laboratory manipulations.

It is therefore a nonnatural biological organism whose genome has been voluntarily modified by man. It is therefore biologically a GMO. However, the modifications carried out to obtain it are not covered by the directive 2001/18/CE. Renan
wheat, the result of laboratory work which has nothing natural about it, is therefore
not considered as a GMO in the regulatory sense of the term.

Box 6.1: Obtaining Renan Wheat
André Gallais describes it in these terms:

The transfer of resistance to eyespot in common wheat was achieved through a t species, durum wheat (*Triticum persicum*), crossing with the two species *Triticum aestivum* and *Aegilops ventricosa*. First, *A. ventricosa* was crossed with durum wheat, then the resulting interspecific hybrid was treated with colchicine to double its number of chromosomes in order to restore its fertility, and then this hybrid was crossed with common wheat, followed by several cycles of crosses with common wheat and selection to maintain resistance to eyespot. The result is a soft wheat progenitor, resistant to eyespot, at the origin of the variety "Renan" [5].

This variety, which has remarkable properties of cold tolerance and resistance to viruses and insects [6], is widely cultivated as soft winter wheat in conventional agriculture but also especially in organic agriculture. However, the specifications of this type of agriculture forbid the use of GMO varieties. But Renan wheat is not considered as GMO, isn't it? Why should organic agriculture deprive itself of using this nonnatural variety but not subject to the opprobrium of a GMO classification?

Conclusion

This comparison between the different definitions of a GMO underlines that it is above all a regulatory definition. It underlines, beyond the divergences on the appreciation of the innovative character which is judged in one case as positive (Canada) and in the other as a source of concern and therefore a priori negative (European Union), that the term regulated GMO is reserved for transgenesis. It also emerges that organisms genetically transformed by random mutagenesis are not "hidden GMOs" but that they are not concerned by the European GMO regulation.

The regulatory definition of a GMO cannot be superimposed on its biological definition, as André Gallais has brilliantly demonstrated.

These considerations lead to some semantic adjustments. Products obtained by techniques dating back more than 30 years are considered today as first-generation biotechnological products. They will be referred to as GMOs. They will be distinguished from second-generation products, resulting from NGTs, which will be qualified as genome-editing products. They could also be called GEO (*genetically edited organisms*), as suggested in a previous book [7]. But let's bet that the acronym that will be retained will be the one recently proposed by the European Commission: NGT products!

Bibliography

1. Rifkin, J. (1998). *Le siècle biotech* (p. 348). French edition by Editions Boréal (Canada).
2. Procassia, C., & Prud'homme, L. (2021). *Les nouvelles techniques de sélection végétale en 2021: avantages, limites, acceptabilité* (p. 122), Rapport n°4220 AN et n°671 Sénat, OPECST, 3 juin 2021.
3. GMO.gouv.pc.ca. (2021). *General information.* http://www.ogm.gouv.qc.ca/information_ generale/info_ogm/info_what.html
4. *Canadian Food Inspection Agency* (CFIA) Agence canadienne d'inspection des aliments (2017). *Novelty and Plants with Novel Traits*, https://inspection.canada.ca/varietes-vegetales/ vegetaux-a-caracteresnouveaux/grandpublic/nouveaute/fra/1338181110010/1338181243773
5. Gallais, A. (2021). Qu'est-ce qu'une variété OGM? Exemple du blé tendre Renan. *Phytoma, 742,* 46–48.
6. Arvalis. (2022). *Fiche variété RENAN Blé tendre d'hiver*, http://www.fiches.arvalis-infos
7. Regnault-Roger, C., Houdebine, L. M., & Ricroch, A. (dir) (2018). *Au-delà des OGM Beyond GMOs* (p. 213). Presses des Mines.

Chapter 7
Is the European Regulation on GMOs Still Justified?

The products of transgenesis (regulated GMOs) are subject in the European Union to the provisions of Directive 2001/18/EC, on which the authorities base the granting of the required approvals.

Before specifying what the European regulations on GMOs are, it is useful to recall what a regulatory approach implies and why it is necessary. We will also ask ourselves whether, in view of the scientific advances made over the last 20 years, the European regulatory requirements are proportionate to the risks involved.

Regulation, Innovation, and the Precautionary Principle

The purpose of a regulation is to establish a legal framework to ensure that the rights of each person defined in the framework of a societal norm are respected.

Innovation, in essence, brings new elements of knowledge and causes changes within a defined framework. To be successful, it must be in line with the expectations of the society that welcomes it. An innovation that responds to a need and offers advantages over what existed previously will be welcomed, even if the extent of the changes that will occur later is not fully understood. In the opposite case, an innovation is doomed to failure if it is misunderstood, and the improvements it proposes will only be accepted if they are presented differently or in a different context. The case of genomic modifications by genetic engineering illustrates this reflection perfectly.

But before developing this example further, it must be emphasized that innovation always involves a degree of uncertainty about future developments and that consequently innovation can be accompanied by positive or negative effects. It is therefore up to those who are responsible for promoting beneficial innovation but also to the public authorities, guarantors in a democracy of collective interests and

© The Author(s), under exclusive license to Springer Nature Switzerland AG 2023
C. Regnault-Roger, *Biotech Challenges*,
https://doi.org/10.1007/978-3-031-38237-6_7

the common good, to reduce the negative effects if they exist, i.e., to allow people to benefit from the advantages while limiting the risks.

In the field of health, food, and the environment, regulations are all the more necessary when a situation is subject to risk, i.e., when it has been noted that harmful effects could occur for the populations to be protected. The regulation must provide for provisions to better protect against a danger if it is present by minimizing it, i.e., by limiting the exposure to this danger and consequently the risk incurred.

Indeed, one should not confuse danger and risk. Risk is the expression of a probability. It is the combination of a danger and exposure to this danger. Thus, when there is no possibility of exposure to a danger, there is no risk. Let's take an example. You live in France where lions do not exist in the wild. You are not at risk from the lion because it is not normally present where you live. There is no natural possibility of exposure to this hazard. But if you live in the African savannah where lions are endemic, you are at risk of encountering a lion, i.e., there is a probability that you will face this danger by unexpectedly encountering a lion.

To reduce the risk, one must minimize exposure to a hazard. However, there must be a probability that a known but virtual danger will become reality. Who does not know the aphorism "an absence of proof is not a proof of absence?" Where to place the cursor of uncertainty? The confusion between danger and risk not only generates a feeling of anxiety or fear but often affects innovation when one wants to have absolute certainty about an absence of risk.

A regulation must therefore foresee and take measures to prevent a plausible and real danger. To do this, it must manage the risks, i.e., the probability of exposure to this danger, by proposing well-thought-out and proportionate decisions based on objective, measurable criteria.

At what point do we assess the imminence of the occurrence of a danger and according to what criteria? From a feeling or an intuitive perception that necessarily has a subjective element or based on parameterized indicators that translate a tangible and verifiable modification of the existing situation?

One might think that a rational legislator, in charge of managing the public space, would unhesitatingly choose the second option. However, this is not what the precautionary principle indicates (Box 7.1).

Box 7.1: The Precautionary Principle and the European Union
The precautionary principle, which originally appeared in international texts of some Northern European countries in the years 1984–1987, was integrated into the principles of the European Treaty of Maastricht and acquired an international scope at the "Earth Summit" in Rio, both events occurring in 1992.

It states that protective measures must be taken to prevent potential or actual risks associated with existing or new situations.

What the European Court has formulated is as follows:

> [...] Where uncertainties remain as to the existence or extent of risks to human health, protective measures may be taken without waiting for the reality and seriousness of those risks to be fully demonstrated.

(continued)

Box 7.1 (continued)
The European Union has incorporated the precautionary principle "as a general customary rule of international law or at least as a general principle of law," but this is not the case in all countries: for example, the United States and Canada do not give this principle the same legal value [1].

The precautionary principle therefore introduces into the regulatory space the notion that it is not necessary to have evidence of the reality of an effect in order to prohibit on the basis of suspicion, without having to demonstrate that there is a certainty that the danger may occur.

In this case, the legislator has opened the door to imaginations filled with anguish and anxiety, to an imagination that is afraid of tomorrow. Humanity has progressed by constantly innovating, improving what exists, and learning from its mistakes. The immobilism by fear of possible unfavorable consequences will prove in the long run to be more prejudicial than the adventure of the innovation.

It is in this context of lack of knowledge on the consequences of biotechnological modifications of the genome, on the one hand, but also of mistrust toward innovation, on the other hand, that the European regulation on GMOs was elaborated in the 1990s.

The European Regulation on GMOs

It is based on a body of texts published over a period of 30 years. A first set of two directives linked together was published in 1989 and 1990, Directives 89/219/EEC and 90/220/EEC on the use of GMOs in contained and open environments, followed 10 years later, in 2001, by Directive 2001/18/EC "on the deliberate release of GMOs into the environment." This is still in force but was amended in 2015 by Directive (EU) 2015/412 on the societal acceptability of these technologies (see Chap. 5). Finally, in 2018, Directive 2018/350/EC updates the regulatory framework for environmental risk assessment monitoring provisions. Regulation (EU) 2015/2283 on novel foods accompanies these directives.

In France, until December 2021, the National Agency for Food, Environmental and Occupational Health Safety (ANSES) for toxicology and the High Council for Biotechnology (HCB) for the environment will analyze the dossiers. These advisory bodies communicate their conclusions to EFSA. The European agency examines them before transmitting its opinion to the European Commission, which will decide.

This regulation induces very detailed files, based on:

- A technical file including a complete description of the plant and its transformation and including a complete toxicological file (alimentarity, allergenicity, etc.).
- An *environmental risk assessment* (ERA) plan specifying exposure and the existence of gene flow or persistence and invasion phenomena and possible immediate, delayed, or cumulative long-term effects on target and nontarget organisms

(fauna, flora of the environment, operators) and on biogeochemical cycles. It should include algorithmic analyses of unlikely future assumptions using theoretical scenarios. It should also indicate what measures are being taken to manage these real or hypothetical risks.

• A *post-market environmental monitoring* (PMEM). This monitoring is conducted over the 10-year authorization period and gives rise to annual reports to the European Commission. PMEM includes general and specific monitoring:

  • General surveillance aims to identify unpredictable changes and unknown unintended effects on nontarget populations that have not been identified as potential targets: it is a "non-hypothesis-driven" approach.
  • Specific monitoring is intended to identify the occurrence of expected changes and to test possible hypotheses about negative effects that may be suspected.

This post-marketing surveillance is very expensive to detect an uncertain and unknown anomaly that would result from the sole cultivation of a transgenic plant.

In the European Union, only one GM crop exists to date, that of the transgenic maize Bt MON 810. Box 7.2 provides an overview of the post-marketing surveillance that has been implemented for almost 20 years on this crop.

Box 7.2: Measures Implemented for the Post-marketing Surveillance Plan for MON 810 Maize in the European Union

Bt MON 810 maize has been cultivated in Spain since 2004. The companies Monsanto and later Bayer after the merger with Monsanto by this company in 2017 are responsible for carrying out the post-marketing surveillance required by Directive 2001/18/EC, which is complemented by Regulations (EC) 1829/2003 and 1830/2003. It is the companies that have requested the approval of the traits (transgenic event) that are responsible for the regulatory file and not the seed companies that market the varieties that have incorporated the GMO modification.

They must carry out a general and a specific monitoring plan when the transgenic crop is cultivated and a general monitoring plan only when it is not cultivated:

• *General monitoring* consists of surveying farmers who grow transgenic maize by means of questionnaires that they fill out and interviews conducted in person or by telephone by experienced independent investigators (generally with more than 10 years' experience in the field): for example, in Spain by *the Instituto Markin*, which specializes in market and opinion research in the agriculture, livestock and agri-food sector; in Portugal by *AGRO.GES*, a research and project company; and in the Czech Republic and Slovakia, when MON 810 corn was grown there, by the Czech University of Agriculture in Prague.

Questions are asked about crop management to identify any anomalies. The areas surveyed must be significant in relation to the total cultivated area (about 10%) and located in all the territories where the crop is grown.

(continued)

Box 7.2 (continued)

Technology User Guides (TUGs) have been distributed to farmers, and local meetings are held to inform them of the specificity of the crop and to provide an annual review.

Those who have questions on a particular point can also contact interactive websites at any time.

The worldwide scientific literature published during the year is also monitored using keywords, and relevant articles are analyzed in depth. In recent years, this monitoring has been carried out by consultancy firms specializing in monitoring and literature analysis and with a solid worldwide reputation.

Naturalist monitoring networks (e.g., the League for the Protection of Birds, hunters' federations, the National Office for Hunting and Wildlife, federations of conservatories of natural areas, etc.) are also mobilized to monitor any anomalies that may occur in regions where GMO corn fields exist.

- *Specific monitoring* is for unintended effects that would be predictable. For Bt maize, what is predictable and needs to be monitored is the development of resistance in target insects (see Chap. 11). Independent research organizations have thus been mobilized to establish baselines of susceptibility to the Bt toxin in target insects (European corn borer and pink stem borer), and they are sampling season after season to check for stability.

In the 2010s, when GMO crops were grown in about ten EU member states, an organization called the *European Union Working Group on Insect Resistance Management* was created by the four international companies (Monsanto, Syngenta Seeds, Pioneer Hi bred Int Incorp, and Dow AgroSciences) that were marketing GMO seeds in Europe at the time. It was in charge of developing a harmonized biomonitoring approach.

At the same time, fieldwork is being carried out. Farmers growing MON 810 are being informed of the refuge zone strategy through meetings, seminars, and leaflets (see Chap. 11). In Portugal, the Portuguese authorities carry out inspections to check that the recommendations are being followed. The overwhelming majority of farmers follow the recommendations of the Technology User Guide (TUG), and the declarations are sincere and do not justify any sanction. In Spain, the Spanish Ministry in charge of GMOs (El MITECO) publishes regular updates.

- *The monitoring of imports of* GMO biological material into the European Union is carried out as part of a general surveillance by *CropLife Europe*, formerly the *European Crop Protection Association,* which succeeded *EuropaBio*. It is an organization at the European level that brings together 22 companies and 32 associations of the crop protection sector established in the Member States. It is responsible for coordinating the general surveillance actions required by Directive 2001/18/EC and Regulation (EC) No 1829/2003.

(continued)

Box 7.2 (continued)

This surveillance is carried out by networks of operators specialized in the import, handling, and transport of corn grain but also of other crops. Three networks are involved:

- COCERAL, network of distributors, wholesalers, importers, exporters dealing with cereals, rice, oilseeds, oils and fats, and livestock feed and more generally in charge of a logistic chain in the food industry.
- Unistock, which brings together professionals in the storage and whole-sale marketing of agricultural products from 12 European countries, is itself part of the COCERAL network.
- FEDIOL, the federation of industrialists producing oils and protein cakes for food use. It brings together 85% of European producers processing more than 150 products based on vegetable oils and fats.

These networks are best able to report potential damaging events from the routine monitoring they perform. They have to:

1. Verify the relevance of the results obtained and to change, if necessary, the choice of methodologies and indicators retained by the HACCP (*Hazard Analysis of Critical Control* Point) method which was chosen to detect anomalies.
2. Provide information to the monitoring networks on the results obtained in the previous year and to report any incident or adverse effect observed.
3. Report annually to the appropriate authorities and to report promptly any adverse effect that has occurred.

It is therefore after a rigorous examination of these huge files on the absence of toxicity for human and animal health or ecotoxicity for ecosystems, and more generally of harmful effects for the environment, that the opinions of the European advisory bodies are issued.

This careful examination of the dissemination in the environment concerns the files for the cultivation of genetically modified plants and the importation of GMO plant productions but also GMO health products, vaccines, and medicines for humans and animals.

Onerous and Discouraging Regulations

In this context, only the large international conglomerates in the sector (see Chap. 12) have the financial basis to take on such regulatory requirements, which are in addition to the normal procedure for applying for marketing authorization (MA) for a new plant variety. A simple application to import a GMO into the EU costs a company between €11 and €16.7 million (a 50% increase in 10 years), and the average time required to study the dossier is 6 years [2]. Small- and medium-sized European companies in the seed sector do not have the means to cope with this.

This body of EU regulations, whose requirements have increased over the years, is therefore not unrelated to the current situation. In addition to the real regulatory obstacle course to which a company wishing to import seeds from a transgenic crop into Europe is subjected, a majority of Member States rely on the mention of "a negative public opinion" to justify refusals to authorize the use of GMOs (crops or imports) on their territories. Faced with this situation, the international groups in the sector are turning away from the European Union and prefer to respond to the demands for new varieties adapted to the climatic and agricultural needs (e.g., resistance to specific diseases or pests) of Asian, South American, and African countries.

Are Regulations Still Relevant?

While it may have seemed appropriate to have such a heavy regulatory framework at a time when there were many unknowns about the behavior of genetically modified plants in the field, is this still the case today?

There are several arguments in favor of reducing the burden and readjusting these regulations:

- The three US academies, the *National Academies of Science, Engineering, and Medicine*, published a report in 2016 entitled *Genetically Engineered Crops: Experiences and Prospects* [3]. This 600+ page report was written after hearing from 80 stakeholders, collecting 700 comments, and analyzing more than 1000 scientific publications related to genetically engineered crops over a 20-year period. After examining the agronomic and environmental effects, the effects on public health, and the social and economic consequences, the three academies concluded that these biotech plants grown according to good agricultural practices do not present any more toxicity and ecotoxicity or environmental risks than conventional plants.
- Due to the scientific advances made in recent years, it is noted that the European regulatory definition of a GMO must be reviewed.
- The environmental risk assessment dossier now seems disproportionate, especially as it has been made more cumbersome over the last 5 years by the obligation to include algorithms for extreme prospective hypotheses. If it was justified at the end of the 1990s when we did not have enough hindsight on these first-generation biotechnologies, is it still relevant in 2022 when there is almost 30 years of experience on the subject?

Contrary to popular belief, the appearance of a GMO field is identical to that of a conventional crop, but it is grown with plant varieties with improved properties. The GM plant can be distinguished in a field by the fact that it is more resistant to diseases and pests and has a healthy appearance, whereas the non-GM plant bears the stigma of pest or pathogen attack (Fig. 7.1). Monitoring of transgenic plant fields is mandatory, as just discussed, as part of post-market surveillance throughout the duration of the crop. While monitoring predictable effects and their occurrence, in order to limit adverse effects, is consistent with good agricultural practices,

Fig. 7.1 Maize in the field in Spain (**a**, transgenic maize MON 810; **b**, conventional maize; **c**, experimental trial; **d**, damaged by the European corn borer *Ostrinia nubilalis*). (Credits Catherine Regnault-Roger)

monitoring for unknown unintended effects on nontarget ecosystems or populations not identified as potential targets is a challenge. However, this is what companies that market seeds classified as GMOs are obliged and committed to do, even though numerous research studies carried out worldwide by independent institutions have now highlighted the absence of adverse effects observed by this general monitoring.

On the other hand, it has also been noted that growing a transgenic crop does not dispense with the need to observe good agronomic practices in order to limit or delay the appearance of foreseeable unintended effects: for example, the strategy of refuge zones, which will be detailed in Chap. 11. Agronomic and educational support on the technical itineraries to be observed is provided to farmers growing transgenic plants by specialized technicians in order to make them aware of the foreseeable risks.

The whole approach underlines that the greatest precautions are observed to limit the unintended effects of GMO crops when they are foreseeable. If specific monitoring seems justified, should we nevertheless continue the general monitoring of GMO crops as it is conceived in the current European regulations, and whose confusing character has been underlined?

Conclusion

It appears today that the European regulations applied to genetically modified plants have become inadequate due to the scientific and technical advances made over the last 20 years. Imbued with the precautionary principle, which one may wonder if it is not rather a principle of inaction and a brake on innovation in certain circumstances, it ignores laboratory and field research and the results obtained: it is, at best, out of date, at worst unsuitable because obsolete. This is why it is urgent to completely overhaul the European regulatory framework applied to plant biotechnologies. The current reflection on the regulatory status of NGTs can create this opportunity.

Bibliography

1. Atlas Magazine. (2010). *A quoi sert le principe de précaution?* Atlas magazine, archives- dossier-focus-world, https://www.atlas-mag.net/article/a-quoi-serves-the-precautionary-principle
2. GMOinfo/en. (2019). *Commerce & authorisations* issue of 28/2/2019, https://www.gmoinfo.eu/fr/articles.php?article=Des-demandes-d%2D%2Dautorisation-d%2D%2DOGM-always-more-expensive-in-Europe
3. Committee on Genetically Engineered Crops: Past Experience and Future Prospects; Board on Agriculture and Natural Resources; Division on Earth and Life Studies of National Academies of Sciences, Engineering, and Medicine (2016). *Genetically Engineered Crops: Experiences and Prospects*, report 420 pages, National Academic Press, https://doi.org/10.17226/23395., http://nas-sites.org/ge-crops/2016/04/27/report-release/

Chapter 8
What Regulation for NGTs in the EU?

Should the regulations applied to first-generation GMOs in the European Union be applied to NGTs? This regulation, undoubtedly one of the most demanding and restrictive in the world, has strongly limited the development of transgenic plants in this part of the world. Today, faced with the generalization of commercial circuits of goods and food freight, the regulation that will be granted to agricultural products derived from NGTs has become a major political issue for the European Union as well as a strategic development choice.

However, the Court of Justice of the European Union (CJEU) ruling of July 25, 2018, has created a new legal situation for biotech products, as it indicates that not only first-generation products but also second-generation products must be subject to Directive 2001/18/EC. This court decision places the European Union in a singular position that not only hinders the development of biotechnology within the European Union but also creates difficulties in the international trade of biotechnology products.

It is therefore giving rise to a profound reflection that aims, beyond the regulatory status to be attributed to second-generation biotechnology products, at a complete revision of Directive 2001/18/EC on which the regulation of GMOs in Europe is based. What are the terms of the current debate in the EU?

The CJEU Ruling: Context and Consequences

Following legal action by nine French local and national anti-GMO environmental associations, the European Court of Justice was asked to give a preliminary ruling on "hidden GMOs" (see Chap. 5) and on the status of products derived from NGTs.

Indeed, to the strategy of organized vandalism in the fields, storage facilities, or laboratories, the opponents of GMOs have added in recent years a more legal tactic mentioned above (Chap. 5), which is accompanied by intense lobbying of certain public authorities.

© The Author(s), under exclusive license to Springer Nature
Switzerland AG 2023
C. Regnault-Roger, *Biotech Challenges*,
https://doi.org/10.1007/978-3-031-38237-6_8

These environmental associations (*Confédération Paysanne, Réseau Semences Paysannes, Amis de la Terre, Vigilance OGM et pesticides 16, Vigilance OG2M, CSFV49, OGM Dangers, Vigilance OGM 33, Fédération Nature & Progrès*) petitioned the French Council of State (*Conseil d'État*) in December 2015 to request a moratorium prohibiting the cultivation in France of herbicide-tolerant varieties (HTVs) obtained through the technique of induced mutagenesis, arguing that they "poison the soil, water and our food" [1]. They thus requested that the regulation that exempts them from being subject to Directive 2001/18/EC be applied against them. After investigating the case, the *Conseil d'État* turned to the Court of Justice of the European Union (CJEU) to ask four preliminary questions on the legal status of these VHT varieties and, by extension, that of NGTs (Box 8.1).

Box 8.1: Preliminary Questions Addressed to the Court of Justice of the European Union (CJEU) by the French Council of State in October 2016 [2]

- "Do organisms obtained by mutagenesis (in particular those obtained by new techniques of directed mutagenesis such as, for example, oligonucleotide-directed mutagenesis (ODM), or directed nuclease mutagenesis (SDN1), which uses different types of proteins (zinc finger nucleases, TALEN, CRISPR/Cas9...) constitute GMOs subject to the rules laid down by the Directive of 12 March 2001? Or should we consider that these organisms obtained by mutagenesis, or only some of them (those obtained by conventional methods of mutagenesis existing before the adoption of the directive) are exempt from the precautionary measures, impact assessment and traceability provided for by this directive?
- Are varieties obtained by mutagenesis 'genetically modified varieties', subject to the rules laid down in the Directive of June 13, 2002, or are they exempt from the obligations laid down in that Directive for the inclusion of genetically modified varieties in the common catalogue of agricultural plant species?
- If the Directive of 12 March 2001 excludes organisms obtained by mutagenesis from its scope, does this mean that the Member States are prohibited from subjecting these organisms obtained by mutagenesis to all or part of the obligations laid down by the Directive or to any other obligation, or do they, on the contrary, have a margin of discretion in defining the regime that may be applied to organisms obtained by mutagenesis?
- If the Directive of 12 March 2001 exempts organisms obtained by mutagenesis from precautionary measures, impact assessment and traceability, can its validity with regard to the precautionary principle (guaranteed by Article 191–2 of the Treaty on the Functioning of the European Union) be questioned? And should we take into account, in this respect, the evolution of genetic engineering processes, the appearance of plant varieties obtained thanks to these techniques and the current scientific uncertainties on their impacts and on the potential risks resulting from them for the environment and human and animal health?"

The CJEU answered these preliminary questions in a judgment dated July 25, 2018. It invoked the precautionary principle and ruled that products obtained by techniques after Directive 2001/18/EC must be subject to this GMO regulation. Those that were exempted by Annex 1 of the Directive could be subject to it, each of the Member States of the Union being free to make its own assessment. This legal position strengthens the regulation of genetically modified products in the EU.

Subsequently, this CJEU ruling was transposed into French law by the Council of State on February 7, 2020, with a very restrictive interpretation since it calls for varieties that have been cultivated for several years but improved by in vitro mutagenesis in the laboratory from spontaneous mutants observed in the field and which are regularly listed in the *Official Catalogue of Cultivated Plant Species and Varieties*, to be removed from the French Catalogue in order to be subject to the regulatory provisions on GMOs. This decision of the French Council of State was however contested by the European Commission which, supported by five Member States (Denmark, Spain, Italy, the Netherlands, and the Czech Republic), published a detailed opinion on August 7, 2020, challenging the conclusions of the French public institution. This detailed opinion is in line with the one issued by the Scientific Committee of the High Council for Biotechnology on June 29, 2020, which indicated, after an in-depth study, that there was no scientific reason to make a distinction between in vitro and in vivo mutagenesis or spontaneous mutations [3]. As a result, the European Commission has invalidated the decisions of the French Council of State, asking it to comply with European regulations, which are part of the single market [4].

Opening of a Public Debate in the EU

This debate has now been going on for 5 years, and there have been several initiatives in response to the CJEU ruling.

A Statement from the Scientific Advice Mechanism

As soon as the CJEU ruling was published, the Group of Senior Scientific Advisors to the European Commission (or SAM (*Scientific Advice Mechanism*)) set to work to analyze it, as the consequences of this legal opinion appeared to be major for the prospects of the scientific and economic development of the European Union.

The SAM is a high-level committee of experts, created on June 9, 2015. It is responsible for independently and transparently informing the European Commission on scientific matters so that it can make informed policy for the Union. Its scientific advice to the Commission must bring together "evidence and ideas from different disciplines, be independent of institutional or political interests, and must take into account the full range of EU policies [5]."

Thus, this high-level expert committee promptly issued a statement on November 13, 2018, entitled *A Scientific Perspective on the Regulatory Status of Genome Editing Derivatives and Its Implications for the GMO Directive* [6]. It analyzes the issues and questions arising from the Court's ruling and the application of the GMO Directive. The SAM points out that "new scientific knowledge and recent technical developments have made the GMO Directive no longer fit for purpose." This text, remarkable for its precision, evokes the evolutions that have taken place since 2001 and underlines that there are difficulties in establishing controls and in realizing the traceability of the products obtained by NGT in commercial exchanges since certain modifications carried out by edition of the genome are a posteriori undetectable. Therefore, the SAM asks that the characteristics of the final product be evaluated instead of legislating based on the method of obtaining it. It insists on the need to consider "current knowledge and scientific evidence, particularly on genome editing and established techniques of genetic modification" and to create a regulatory environment favorable to innovation so that "society can benefit from new science and technology." It also calls for societal dialogue among all stakeholders.

The Grow Scientific Progress: Crops Matter!

A European citizens' initiative *Grow Scientific Progress: Crops Matter*! was launched on July 25, 2019, by a group of students from Wageningen University (*Wageningen Universiteit en Research Centrum*, WUR) of eight different nationalities. It called for a change in the current legislation:

> To take into account important advancements in plant breeding techniques. Most importantly, we propose to focus on the crop rather than the technique. In this way safety is ensured while the valuable benefits of new techniques are not lost to illogical regulatory hurdles [7].

This European petition was closed on July 25, 2021, and the number of signatories was disappointing, as it did not exceed 10,000. The most represented countries among the signatories of the petition were, in descending order, Germany (2245), Italy (2188), Spain (995), France (825), the Netherlands (610), Sweden (591), Belgium (427), and Portugal (385). In relation to the population of the EU Member States, the highest percentages were obtained in Italy (4%), Sweden (3.94%), the Netherlands (3.13%), Germany (3.12%), Belgium (2.71%), Spain (2.46%), and Portugal (2.44%), the last two of which grow GMO maize to a significant degree on their national territory. France ranks behind Austria and the Czech Republic!

This nice initiative of students, who are the future of Europe, is considered a failure. Is it a lack of communication? Refusal of biotechnologies or lack of interest from citizens? The mobilization on the subject was not at the appointment, whereas the objectives were praiseworthy (Box 8.2).

Box 8.2: The Objectives of the Citizens' Initiative "Grow Scientific Progress: Crops Matter!" by European students at Wageningen University (2019–2021) [8]

"Directive 2001/18/EC is outdated. We as students of Life Sciences consider it important to revise the current exemption mechanism stipulated in the Directive, with regards to new plant breeding techniques (NPBTs).

Specifically we demand for more legal clarity and further definitions. Our objective is to facilitate the authorization procedure for those products obtained through NPBTs, which carry only natural existing traits. The aim is to improve scientific progress in the European Union while protecting human and animal health and the environment."

Elected Representatives of the German Green Party (Grünen)

Political parties also spoke out. For example, several prominent members of the German Green Party (*Grünen*), who are elected to various important political bodies such as the *Bundestag* (German parliament) and the Senate of the city of Hamburg (among them Katharina Fegebank, Anna Christmann, and Kai Gehring), published a manifesto in June 2020 with the title *Neue Zeiten, neue Antworten: Gentechnikrecht zeitgemäß regulieren* [New times, new answers: regulating genetic engineering law in a modern way].

They point out that genetic engineering applied to human health is universally accepted and that applications in agriculture can also be made within the framework of sustainability "with the right framework conditions" to save time in the face of future challenges like climate change. Noting that the current regulation favors "monopoly structures in agriculture" and hinders public research, they insist on the need for new rules to give public institutions and medium-sized companies an opportunity to use these new techniques to better respond to innovation challenges. They conclude that European regulations no longer correspond to the current state of science and call for the implementation of "a balanced and prudent technology assessment in dialogue with science," considering the result obtained and not the technology used [9].

Since the end of November 2021, the German Greens have been participating in their country's government. Will the supporters of this open position on biotechnology, including some of the party's leading figures, be able to make their voice heard in favor of progress in this field?

Academic and Parliamentary Institutions

In November 2020, it was the turn of the *Union of European Academies of Agriculture* (UEAA), supported by several academies of Member States, to express concern about the fact that 80% of patents filed on applications of the CRISPR/Cas technique are American or Chinese and less than 10% European (cf. Chap. 13). It has taken a stand to demand new regulations on GMOs adapted to modern techniques of varietal selection [10]. Then, in a press release issued on May 4, 2021, 4 days after the European Commission's publication of April 29, 2021, it deplored the fact that insufficient attention had been paid to the progress that genome editing allows in terms of animal welfare and veterinary public health (cf. Chap. 15). Following the work of a committee of international experts from seven EU Member States that it convened, the UEAA adopted recommendations for:

> a novel approach regarding the governance of Gene Edited plants and animals and a new regulation frame adapted to most recent scientific advances [11].

In June 2021, the Parliamentary Office OPECST, which brings together senators and deputies to enlighten the French Parliament on current scientific issues, issued a report (mentioned in Chap. 6) signed by Senator Catherine Procaccia as main author [12], extending the work carried out during the Office's previous term [13]. This report follows a public hearing held on March 18, 2021, under the direction of the two parliamentarians. It insists on the need to revise Directive 2001/18/EC and recommends that the risk assessments of new products be based on their final characteristics and not on the production technique. He proposes that the directive be revised every 5 years to consider technical advances and public debate. He denounces the anti-biotech vandalism and worries about the brain drain abroad as well as the difficulties to conduct biotechnology research in France. He expresses the concern not to obstruct the research on NGT in the EU, both in laboratory and in "real cultivation conditions, without fear of crop destruction." He also wants the government to take a position in the dispute between the French Council of State and the European Commission regarding the application in French law of the CJEU ruling of July 25, 2018 (see Chap. 5).

Opening of a European Strategic Initiative

It is in this context that the European Commission, in accordance with the decision of the Council of the European Union of November 8, 2019, has mandated the JRC to draw up a state of the art on NGTs, inviting it to submit a study on the status of new genomic techniques in Union law accompanied, if necessary, by proposals. On the basis of these conclusions, the Commission published a letter on April 29, 2021, instructing the country holding the rotating presidency of the European Union to make proposals for changes to the legal framework for NGTs within the Union and to organize a debate on the subject. In its message, it insists on the fact that we must first focus on plant applications [14]. Doesn't the improvement of animal health and welfare of farm animals also deserve to benefit from NGTs [15]?

The review process is complex and involves several steps: First, the European Union's scientific and technical research laboratory (JRC (*Joint Research Centre*)) is responsible for drawing up a state of the art on NGTs and the progress of R&D projects. The JRC submitted two reports in spring 2021 [16, 17], and the impact assessment phase, which is open to the public, was initiated in September–October 2021. The latter was the target of a campaign by MEPs from the Greens/EFA (European Free Alliance) parliamentary group entitled "Keep GMOs out of our fields and plates," accompanied by a cyberattack that flooded the consultation with 69,000 spam messages. The maneuver having been foiled, the results of this stage allowed the process to continue. The next step, a public consultation entitled "Legislation for plants produced by certain new genomic techniques," took place between April and July 2022, and its results were published in September 2022. Of the 2300 validated responses, four-fifths (80%) were in favor of revising the EU regulations, while less than one-fifth (17%) supported maintaining the current regulations. Based on these results, the regulatory review process will be completed, and the European Commission will propose a regulatory amendment to the European Parliament, the terms of which it will specify in the summer of 2023 [18].

Conclusion

The European Commission noted in its comments of April 29, 2021 [19], that "there is considerable interest for NGT-related research in the EU. For many EU countries and stakeholders, the current regulatory framework has a negative impact on EU public and private research and innovation in NGTs." It insisted on the potential of NGTs for the objectives of sustainable agriculture that the Union has set for itself through the *Green Deal* approach and the complementary strategies of *Farm to Fork (F2F)* and *Biodiversity*. It also uhighlighted the trade and regulatory impediments thaut the CJEU ruling of July 25, 2018, was creating. This is detrimental primarily to European small- and medium-sized enterprises (SMEs) engaged in the second-generation biotechnology sector.

It is indeed all these aspects that should be taken into account because it is indeed the agroindependence of the EU in a globalized world that is at stake.

What are other countries doing? What regulatory approaches to NGTs have they chosen? This is what Chap. 9 is about.

Bibliography

1. Confédération paysanne. (2017). *Communiqué de presse: OGM cachés: 9 États et institutions européennes donnent leurs avis à la Cour de Justice Européenne*, https://www.confederation-paysanne.fr/actu.php?id=5770
2. Conseil d'État et la juridiction administrative. (2016). *Organismes obtenus par mutagenèse*, http://www.conseil-etat.fr/Actualites/Communiques/Organismes-obtenus-par-mutagenese.
3. High Council for Biotechnology. (2020). *Opinion of the HCB on the draft decree modifying article D.531-2 of the environment code*, http://www.hautconseildesbiotechnologies.fr/sites/www.hautconseildesbiotechnologies.fr/files/file_fields/2020/12/11/avis-cs-hcb-projet-decret-modifiant-code-environnement-200707-rev-201203.pdf
4. Regnault-Roger, C. (2021, May 17). Regulatory and political challenges of new breeding techniques. *Europeanseed*, https://european-seed.com/2021/05/regulatory-and-political-challenges-of-new-breeding-techniques/
5. European Commission. (2021). *Group of Chief Scientific Advisors*, https://ec.europa.eu/research/sam/index.cfm?pg=hlg#
6. Group of Chief Scientific Advisors. (2018, November 13). *A scientific perspective on the regulatory status of products derived from gene editing and the implications for the GMO Directive Statement by the Group of Chief Scientific Advisors*, https://op.europa.eu/en/publication-detail/-/publication/a9100d3c-4930-11e9-a8ed-01aa75ed71a1/language-en
7. Regnault-Roger, C. (2020). European Citizens' Initiative: Sign the petition for green biotech. *European Scientist*, https://www.europeanscientist.com/en/agriculture/european-citizens-initiative-sign-the-green-biotech-petition/
8. European Union Website. (2019). *Cultivating scientific progress: Crops matter!* https://europa.eu/citizens-initiative/initiatives/details/2019/000012_fr
9. Regnault-Roger, C. (2020). An initiative of German Greens in favor of new green bio-technologies. *European Scientist*, https://www.europeanscientist.com/en/opinion/an-initiative-of-German-Greens-in-favor-of-new-green-biotechnologies/
10. UEAA Info. (2020). *Position paper Gene editing and New EU Regulations Urgently needed*, https://ueaa.info/wp-content/uploads/2020/11/Gene-editing-and-new-EU-regulations-urgently-needed.pdf
11. UEAA Info. (2022). The UEAA recommendations for an EU regulation frame concerning Genome editing research and development for crop plants and farm animals 3. 01. 2022, https://ueaa.info/
12. Procassia, C., & Prud'homme, L. (2021). *Les nouvelles techniques de sélection végétale en 2021: avantages, limites, acceptabilité* (p. 122). Rapport n°4220 AN et 671 Sénat, OPECST, 3 juin 2021.
13. Le Déaut, J. -Y., & Procaccia, C. (2017). *Les enjeux économiques, environnementaux, sanitaires et éthiques des biotechnologies à la lumière des nouvelles pistes de recherche* – Tome I, report 4818 (National Assembly) and 505 (Senate), of 13 April 2017, https://www.assemblee-national.fr/14/rap-off/i4618-tI.asp
14. European Commission. (2021, April 29). *EC study on new genomic techniques. Letter to the Portuguese Presidency*, https://ec.europa.eu/food/plants/genetically-modified-organisms/new-techniques-biotechnology/ec-study-new-genomic-techniques_en
15. Thibier, M., & Regnault-Roger, C. (2022). Presentation du livre "Enjeux biotechnologiques. Des OGM à l'édition du génome". *Bulletin Académie Vétérinaire France, 175*. https://doi.org/10.3406/bavf.2022.71002
16. Broothaerts, W., Jacchia, S., Angers, A., Petrillo, M., Querci, M., Savini, C., Van den Eede, G., & Emons, H. (2021). *New genomic techniques: State-of-the-art review* EUR 30430 EN. Publications Office of the European Union, Luxembourg. https://doi.org/10.2760/71005 6JRC121847

17. Parisi, C., & Rodríguez-Cerezo, E. (2021). *Current and future market applications of new genomic techniques* EUR 30589 EN. Publications Office of the European Union, Luxembourg. https://doi.org/10.2760/02472JRC123830

18. Regnault-Roger, C. (2023). *French Academy of Agriculture scientist challenges government to 'follow the science' and revise its regulatory opposition to genetically edited crops*, Genetic Literacy Project 2023-03-30, https://geneticliteracyproject.org/2023/03/30/french-academy-of-agriculture-scientist-challenges-government-to-follow-the-science-and-revise-its-regulatory-opposition-to-genetically-edited-crops/

19. European Commission. (2021). *EC study on new genomic techniques. Executive summary,* Commission staff working document of 29.04.21 SWD(2001)92, https://ec.europa.eu/food/plants/genetically-modified-organisms/new-techniques-biotechnology/ec-study-new-genomic-techniques_en

Chapter 9
What Are Regulations on NGTs Elsewhere in the World?

At the same time as the CJEU issued its ruling equating NGTs with GMOs, 13 countries (Argentina, Australia, Brazil, Canada, Colombia, Dominican Republic, Guatemala, Honduras, Jordan, Paraguay, the United States, Uruguay, Vietnam), as well as the Secretariat of the Economic Community of West African States (ECOWAS), sent a manifesto to the WTO (World Trade Organization) in November 2018 stating that:

> support policies that enable agricultural innovation, including genome editing" and also that "Precision biotechnology techniques, as a whole, constitute an essential tool for agricultural innovation. Their use provides farmers with access to products that increase productivity while preserving environmental sustainability. [1]

The manifesto stated that "arbitrary and unjustified distinctions" between crops derived from genome-editing or conventional breeding should be outlawed.

The signatory countries belong to the five continents, thus giving a worldwide echo to this approach.

What regulations on NGTs are currently in place in different parts of the world? We will present here the situation in several countries. However, the positions they have taken can only be understood by considering the history of biotechnology development in the different parts of the world and the state of progress, which will be explained in the second part of this book.

Indeed, if biotechnologies have been well accepted worldwide at the medical level, it is not the same for agricultural applications. This rejection has been particularly directed at plant biotechnology and plant breeding. Chapter 5 reported on the organized defiance of transgenic soybeans imported into Europe. Other campaigns by anti-GMO NGOs have led to divide the world in two: on the one hand, countries that have fully adopted first-generation biotechnologies for cultivation and trade (imports, exports) and countries that refuse cultivation on their territory without refusing trade (see Chap. 12) [2]. This divide is reflected in the way NGTs have been received, with regulatory approaches that are flexible depending on the country.

American Continent

The American continent is the continent whose countries have widely adopted not only first- but also second-generation biotechnologies. A majority of these countries are Latin American. South America has been called the "land of choice" for biotech plants, as four countries have more than half of their arable land cultivated with transgenic plants (Uruguay, Argentina, Paraguay, and Brazil) [3]. It is therefore not surprising that they were pioneers in ruling on NGTs.

In these countries, the marketing of genetically modified products is conditional on the approval of an organization in charge of biosafety, which must respond to the request for authorization within a specific timeframe, and genome-editing products that do not incorporate foreign DNA are generally not considered GMOs.

Thus, four pioneer countries, Argentina (resolution 173/2015), Chile (normative resolution 2017), Colombia (resolution 29299/2018), and Brazil (normative resolution 16/2018), soon followed by Paraguay in 2019, then Ecuador, and in Central America also Honduras, El Salvador, and Guatemala, have decided to proceed on a case-by-case basis while exempting from regulation any new genetically modified organism by NBT/NGT that does not incorporate "new combinations of genetic material."

Argentina adopted the world's first regulation on the status of NGTs in 2015. It entrusted CONABIA (*Comisión Nacional Asesora de Biotecnología Agropecuaria*) with the examination of applications for authorization, which decides on a case-by-case basis according to the techniques used, the genetic modification performed, and whether there is a transgene present in the final product. When a genetically edited variety is exempted from GMO regulations, it is labeled like conventional varieties without any special mention.

Guatemala created a simplified procedure in 2019 to register seeds that do not have DNA additions and have already been approved by countries with trade agreements with it. Neighboring Costa Rica has specific regulations for the cultivation, import, and export of biotech plants but has not adopted specific regulations for the registration of products derived from them for human and animal consumption or processing [4]. The status of genome-editing products is currently under discussion.

In North America, in the United States, new rules, called *SECURE (Sustainable, Ecological, Consistent, Uniform, Responsible and Efficient) Rule,* are applied to the new genome-editing biotechnologies after a vast consultation was organized to collect the opinion of citizens. They were published on May 18, 2020, by the *Federal Register*, the official journal of the United States.

This regulation states that plants genetically edited for minor modifications of the genome such as the change or deletion of a base pair or the introduction of a gene known to belong to the plant's gene pool (modifications of the type SDN1 and SDN2) will be exempted from the federal regulations applied to GMOs [5]. It is the characteristics of the final product that are now evaluated and not the method of obtaining it.

The US Department of Agriculture's (USDA) *Animal and Plant Health Inspection Service* (APHIS)[2] estimates that more than 99% of the new varieties submitted to the USDA for market access will benefit from this regulatory relief. The USDA has found that the current regulatory environment for GMOs creates disincentives for innovation. The USDA hopes that this new procedure will encourage innovation, particularly in plant breeding [6].Canada, another North American country, does not treat genome-edited products differently from other products resulting from innovations that have novel traits (see Chap. 6). The method of production (mutagenesis, transgenesis, NBT, conventional crossing, etc.) does not matter. What matters are the properties of the resulting finished product, which are evaluated on a case-by-case basis by the *Canadian Food Inspection Agency* (CFIA). The CFIA is responsible for the regulation of the environmental release of plants with novel traits and its oversight falls under the *Plant Protection Act* and Regulations and the *Seeds Act* and Regulations.

Asian Continent and Pacific Area

The Asian continent is home to countries with very disparate situations, not only in terms of development but also in terms of economic and technological opening policies.

Japan is a country that is very open to biotechnology, and as early as 2017, it began a process to rule on NGT-modified products. The *Ministry of Health, Labour and Welfare* (MHLW) and the *Ministry of Agriculture, Forestry, and Fisheries* (MAFF) published in 2020 the regulatory guidelines to be applied to genome-editing products. These guidelines state that evaluations are made on a case-by-case basis, which requires the submission of a dossier providing information on the genetic modification performed and the technique used. This application for authorization must be made for the development of the genetically edited plant and for subsequent crosses with conventional plants. There is no health and environmental assessment procedure in the absence of genes foreign to the species. Regional governments have the possibility to make additional regulatory adjustments if they wish. Since 2021, genetically edited tomatoes have been marketed in Japan (see Chap. 14).

China is also a very advanced country for first- or second-generation biotechnologies. It has not defined a specific regulatory status for genome-editing products but has undertaken research programs worth US$10 billion. It rivals the United States in terms of scientific publications on CRISPR and patents on NGTs (see Chap. 13). It is also the country with the most patents for agricultural applications of CRISPR/Cas9 [7].

India, another major country in the region that is developing very dynamically, has adopted a less advanced position than the two previous countries, no doubt due to the presence of very active anti-GMO NGOs of which Vandana Shiva, a world-renowned activist, is the leading figure. While initially genome-edited products were subject to the same regulations as GMOs (*Rules for the Manufacture, Use, Import, Export and Storage of Hazardous Microorganisms, Genetically Engineered Organisms or Cells*), the Indian government revised the rules. In March 30, 2022, the *Ministry of Environment, Forest and Climate Change* announced the exemption of genome-edited plants without exogenous introduced DNA (derived by SDN1 and SDN2) and the *Ministry of Science and Technology* published "Guidelines for the Safety Assessment for the Genome edited Plants, 2002" in May 17, 2022. Genome-edited plants including foreign genes (SDN3-type), are falling under GMO-type dossier.

In the Pacific Zone, Australia has ruled on the regulation of NGTs through a 2019 amendment to the *Gene Technology Regulations* applied since 2001. SDN1 genome-editing products are considered traditional mutation products and are exempt from regulation, while a dossier is required for SDN2 and SDN3 [8].

New Zealand had ruled in 2016, following a 2014 decision by its High Court of Justice, that genome-editing products should be considered GMOs. But following debates led by the *Royal Society Te Apārangi Gene Editing Panel* of New Zealand in the years that followed, a reflection is underway to change the status of certain genome-editing products. Indeed, while stressing that regulation must be proportional to the risks involved, these debates have emphasized that New Zealand's position must be nuanced to consider not only the characteristics of gene-edited products but also the international exchanges that this country maintains with, in particular, its neighbor and economic partner Australia [9].

Several less economically advanced countries in Southeast Asia are still considering how to regulate NGTs.

Other Continents

On the European continent, besides the European Union, two countries are very reserved about first-generation biotechnologies.

Norway, which is not part of the EU, is developing some of the strictest regulations against genetically modified products through transgenesis. But it is moving toward considering that genome-editing products that do not incorporate foreign DNA would not be considered GMOs.

The second country, Russia, recently reaffirmed in 2020 its opposition to the cultivation and breeding of agricultural GMOs except for research purposes. However, it has been developing a 111-billion-ruble (about €1.3 billion) research program since 2019 to develop about 30 genetically edited varieties of wheat, barley, sugar beets, and potatoes (10 varieties in 2020 and then 20 in 2027). The decree issued for this purpose states that these new varieties should be considered equivalent to conventionally bred varieties [8].

In the Mediterranean area, Israel decided in 2016 not to regulate gene products that do not contain DNA from other species. Its *National Committee for Transgenic*

Plants (NCTP), which made this decision, said that other genetic modifications will be considered on a case-by-case basis under the *Seed Act*, which regulates GMOs in Israel.

In Africa, where there are a few research centers working on genome-editing products to address local problems (see Chap. 16), discussions are taking place in Nigeria between stakeholders, civil society organizations, and government authorities to manage genome editing in the context of biosafety and to include it in the regulations. In Kenya, several researchers are calling for clarification on the regulations that will be applied to the products of their research to be able to work with good visibility [10].

Conclusion

What are the consequences of these regulatory adjustments in several countries around the world? It seems that Argentina, which opted early, in 2015, for regulatory relief, is observing that the deregulation is accompanied by an expanded supply of new genetically edited products by SDN1 and SDN2, with new varieties that are more efficient in adapting to climate change or more resistant to pathogens and crop pests [11].

An examination of the regulations adopted by most countries for genome-editing products shows that the plants that carry minor undetectable modifications of the genome (SDN1 and SDN2 type) are exempted from the obligation to be regulated as GMOs in these countries. The more substantial SDN3 modifications, which can be detected, will be subject to the regulations.

Bibliography

1. English, C. (2018, November 7). 3 nations say it's time to end 'political posturing' and embrace crop gene editing. *Genetic Literacy Project*, https://geneticliteracyproject.org/2018/11/07/13-nations-say-time-to-end-political-posturing-and-embrace-crop-gene-editing/
2. ISAAA. (2020). Global status of commercialized biotech/GM crops in 2019. *Brief- 55*, https://www.isaaa.org/resources/publications/briefs/
3. Regnault, H. (2019, November 11). *Three agricultural revolutions: Between technical innovations and social mutations.* Conference at the University of Costa Rica, San José.
4. GAIN Report. (2018, December 12). Costa Rica agricultural biotechnology annual. *Biotechnology Annual Report*, number 18016, p. 9.
5. Juarez, B. (2020). *Sustainable, ecological, consistent, uniform, responsible and efficient (SECURE Rule) overview,* USDA, https://www.aphis.usda.gov/aphis/ourfocus/biotechnology/biotech-rule-revision
6. Regnault-Roger, C. (2020). A new regulation in the United States to promote agricultural innovations. *The European Scientist, 28*(05), 2020.
7. Regnault-Roger, C. (2020). *GMOs and genome-editing products: Regulatory and geopolitical issues* (p. 56). Foundation for Political Innovation.

8. Global Gene Editing Regulation Tracker. (2021). *Australia crops/food,* https://crispr-gene-editing-regs-tracker.geneticliteracyproject.org/australia-crops-food/

9. Royal Society Te Apārangi Gene Editing Panel. (2019). *Gene editing: Legal and regulatory implications* (p. 16), https://www.royalsociety.org.nz/what-we-do/our-expert-advice/all-expert-advice-papers/gene-editing-legal-and-regulatory-implications

10. Aghan, D. (2019, July 15). We need policy on new breeding technologies. *The Standard,* https://www.standardmedia.co.ke/commentary/article/2001333855/we-need-policy-on-new-breeding-technologies

11. Regnault-Roger, C. (2020). *Les biotechnologies agricoles, une clé pour l'indépendance agro- alimentaire,* Institut sapiens, https://www.institutsapiens.fr/les-biotechnologies-agricoles-une-cle-pour-lindependance-agro-alimentaire

Part II
GMOs in the World

Chapter 10
GMOs: Medical and Animal Applications

The use of GMOs for therapeutic purposes is widely accepted to the point that public opinion has forgotten that millions of lives have been saved thanks to biotechnologies.

These advances result from a wide range of approaches to better understand the mechanisms of pathologies or to develop innovative therapeutic solutions and biofortified foods.

Laboratory Animals to Better Understand and Fight Diseases

Genetically modified animals have become essential for understanding disease mechanisms and the role of genes in disease expression.

With the biotechnological tools of mutagenesis or transgenesis, it is possible to study the function and role of genes by modifying a gene, inactivating it or replacing alleles to examine the effects produced and the abnormalities that occur.

Several animal species are used for this purpose. For the study of human diseases, genetically modified mammals are most often used because they are physiologically closer to our human species. They could be modified to mimic a syndrome and find a solution to resist it. These transgenic animals are also used in preclinical studies to evaluate new therapeutic approaches, their efficacy, or their toxicity.

The mouse was a species used very early on because its anatomy and physiological functioning are well known (which student does not remember practical work on dissecting a mouse!) and its breeding is one of the cheapest. The embryonic modification of mice to obtain transgenic animals was carried out as early as the 1980s (see Chap. 2).

A huge number of diseases have been studied using these GMO mice, and one of the latest examples is a supplier in Bar Harbor (Maine, USA) that produced more than 3000 mice genetically modified by transgenesis to synthesize a human protein

C. Regnault-Roger, *Biotech Challenges*,
https://doi.org/10.1007/978-3-031-38237-6_10

involved in the infectious mechanism of coronavirus in the spring of 2020 in order to better understand the COVID-19 pandemic [1].

Although the rat is also popular as an experimental species, especially for behavioral studies, obtaining transgenic rats has proven to be more difficult [2]. Other species used include rabbits and pigs, which are closer to humans (as demonstrated by organ transplantation experiments).

Less expensive than farmed mammals, other species are used as laboratory models to study certain mechanisms, in particular zebra fish (*Danio rerio*), an oviparous fish, vinegar fly (*Drosophila melanogaster*), and silkworm (*Bombyx mori*).

Therapeutic Advances

As early as the 2000s, Louis-Marie Houdebine identified the therapeutic progress brought about by GMOs [3]. The easy development of genetically modified microorganisms that are then cultivated in incubators for the industrial production of human hormones led to the commercialization of therapeutic solutions in as early as the 1980s. Subsequent technical progress made it possible to modify higher organisms to biosynthesize compounds of pharmaceutical use or to improve their nutritional profile.

GMO Microorganisms

This is the oldest approach, and tangible results have been achieved since the early 1980s. Here are some examples:

- The synthesis of human insulin to treat diabetes, which was commercialized in 1982. Previously, diabetics had to resort to pig insulin extracted from the animal's pancreas which, although very similar to human insulin, differs from it by an amino acid and could cause therapeutic accidents. Bacterial human insulin is synthesized by a widespread bacterium, *Escherichia coli*, genetically engineered to include the human gene coding for this hormone. This insulin, which is completely identical to that produced by the human body, considerably reduces the risk of allergy in patients.
- The treatment of dwarfism by growth hormone. For a long time extracted from the pituitary glands of human cadavers, this growth hormone could be infected by various viruses or prions which are abnormal glycoproteins responsible for Creutzfeldt-Jakob disease or subacute spongiform encephalopathy. This pathology is a neurodegenerative disease that causes mental disorders rapidly evolving toward dementia. In animals, it is called scrapie (sheep and goat) or bovine spongiform encephalopathy (cow). This disease was in the medias particularly in the years 1985–2004 in what was called "the mad cow crisis." Since 1988, the

growth hormone prescribed in France is no longer extracted from cadavers but produced by genetic engineering from genetically modified microorganisms, limiting undesirable contaminations.

- The production of EPO or erythropoietin. Well-known in certain sporting circles where doping is widespread, EPO is a natural renal hormone. It acts on the bone marrow to stimulate the production of hemoglobin and red blood cells, thus improving oxygen transport. Its administration is indicated for patients suffering for anemia and kidney failure. Since 1983, genetic engineering methods have made it possible to produce this recombinant human protein[1] in the laboratory, which is now part of the therapeutic arsenal in hospitals.

Vaccine production has also been revolutionized by genetic engineering. For a long time, vaccines were given by administering weakened or killed pathogens or fragments of pathogens or antibodies active against the targeted infectious agent, so that the body could react as soon as the pathogen was invaded. The industrial process for developing these vaccines was complicated and time-consuming. Since 1983, vaccines, such as those against hepatitis B or influenza, smallpox, hepatitis A, poliomyelitis, tetanus, or HPV (against the human papillomavirus, a cancer-causing agent), have been produced from cell cultures or yeast modified by genetic engineering.

More recently, vaccines against COVID are another illustration of the role of genetic engineering. Besides mRNA technology, vaccines have been developed by modifying inactivated or attenuated viruses or plants (tobacco) through genetic engineering. While the former (e.g., Pfizer and Moderna) are not considered GMOs (since mRNA is a nucleic acid but not a GMO), the latter (Janssen, AstraZeneca, Sputnik V vaccines, etc.) are and have gone through the European GMO authorization procedure for their marketing in the EU. Only the first two have been approved by the European Medicines Agency (EMA).

Transgenic Animals

Recombinant proteins (GMO) of pharmaceutical interest are also obtained from genetically modified animals:

- An anticoagulant factor, human antithrombin (III) has been obtained from the milk of transgenic goats since 2006 and is used in certain treatments for thromboembolic events or to prevent venous thrombosis.
- A glycoprotein used in the management of hereditary angioedema (HAE), a rare genetic disease, and a protein with effects against septic shock (*plasma phospholipid transfer protein* (PLPT)) were produced in rabbit milk in 2010 and 2017 [4].

---

[1] Produced by a cell whose DNA has been modified by genetic engineering.

Transgenic Plants

There are also genetically modified plants that produce drug molecules. They are grown in greenhouses or in the open field. We also use plant cells transformed in fermenters.

Thus, Gaucher disease or sphingolipidosis, a rare disease due to an enzyme deficiency, is treated with taliglucerase alpha, a recombinant protein synthesized by transgenic carrot cell cultures [5].

A second example is the production of anti-Ebola antibodies and COVID-19 vaccines from tobacco. *Medicago*, a company based in Quebec City (Canada), and *British American Tobacco*, one of the world's largest tobacco producers through its subsidiary KBP (*Kentucky BioProcessing*), developed recombinant protein-based vaccines and monoclonal antibodies from the tobacco species *Nicotiana benthamiana* [6]. Phase 3 clinical trials for Medicago-GSK's (*GlaxoSmithKline*) SARS-CoV-2 coronavirus vaccine candidate were completed in December 2021 with a 75% efficacy rate [7], and as a result, the vaccine was licensed by *Health Canada* on February 24, 2022, under the name Covifenz. However, the *World Health Organization* (WHO) refused to approve the vaccine, citing the fact that *Philip Morris International* is a minority shareholder (20% of the capital) in Medicago [8], which is a barrier to approval! We should point out to the WHO that *Philip Morris International* is certainly a global cigarette company, but for several years, it has been developing, alongside innovative alternative products to conventional cigarettes to reduce their harmfulness (IQOS heated tobacco process), many research activities for the conversion of tobacco cultivation for biotechnological and therapeutic uses [6]!

Improving Human Nutrition

Some food plants are also "biofortified," i.e., modified to increase their nutritional value in the edible parts (seeds, fruits, tubers, and/or leaves); others are modified to improve the nutritional profile by increasing the content of desirable nutrients (polyunsaturated fatty acids, provitamin A) and decreasing the content of undesirable compounds (transfatty acids, erucic acid, acrylamide).

Biofortified Foods

Several food plants are being "biofortified." For example, the Bill & Melinda Gates Foundation is supporting a transgenic breeding program for an African plantain "matooke" that aims to protect it from pathogenic microorganisms and to increase its provitamin A (β-carotene) content [9]. But the most publicized example is that of "golden rice," a genetically modified rice variety that has been the subject of societal controversy (Box 10.1).

Box 10.1: About Golden Rice
The project to create this transgenic rice (*Oryza sativa*) was born in 1992. Its objective was to transform this plant so that it synthesizes a precursor of vitamin A, β-carotene. Avitaminosis A causes xerophthalmia, a disease that affects, according to the WHO, more than 5 million preschool children in poor populations and nearly 10 million pregnant women in developing countries.

Its authors Ingo Potrykus of the *Institute of Plant Sciences* at the Swiss Federal Institute of Technology in Zurich, Switzerland, and Peter Beyer of the University of Freiburg Im Breisgau, Germany, published in the journal *science* in 2000 the results of their work that genetically engineered a novel biosynthetic pathway for the plant to produce β-carotene. Their early work integrated genes from a narcissus flower and a bacterium *Erwinia uredovora* into rice. But β-carotene production was low and slow. Further work was being done in partnership with the Swiss company Syngenta to achieve a more efficient transformation by replacing the narcissus gene with a corn gene. The β-carotene content was increased 23-fold, and the transformation was only expressed in rice grains that are precisely consumed by the people.

This research was financed by many organizations. No less than 30 companies and universities can claim intellectual property rights to the invention. However, at the instigation of Ingo Potrykus, Peter Beyer, and Syngenta, it was decided that golden rice would be freely licensed for humanitarian use. A legal framework was drawn up for this purpose, so that farmers in developing countries could have access to this seed at no extra cost compared to conventional varieties.

After an intense smear campaign by the NGO Greenpeace and Indian activist Vandana Shiva, golden rice has been plundered, including experimental golden rice trials vandalized in the Philippines in 2014. In response, 107 Nobel laureates signed and sent a manifesto to the UN in 2016 to disavow the Greenpeace campaign, calling it a "crime against humanity."

Since early 2018, Australia and New Zealand and then Canada and the United States have allowed golden rice to be marketed for food consumption to encourage cultivation in tropical countries. And at the end of July 2021, the Philippines became the first country in the world to allow the cultivation of golden rice for commercial purposes. Bangladesh is expected to follow soon. Indonesia, India, and China will join them soon.

Nearly 30 years of research and struggle to impose this culture, the golden GMO rice has become a symbol!

Improved Nutritional Profile

Among the food plants genetically modified to improve the nutritional profile of plants, two examples can be cited:

- Canola, a major GMO crop in Canada, is derived from rapeseed varieties modified to produce a food oil with a low erucic acid content, a fatty acid whose consumption could present a long-term safety risk in children under 10 years of age (myocardial lipidosis observed in animals).
- Ranger Russet, Russet Burbank, and Atlantic potatoes have been modified using an innovative technology to resist internal blackening or browning due to the presence of enzymes that form melanin from the amino acid tyrosine when the plant cells have been damaged. This physiological disorder occurs as a result of shocks that bruise the tuber during harvesting or during the sorting and handling phases.
  A second modification was made in these varieties to reduce the amount of acrylamide contained in varieties rich in asparagine, another cellular amino acid of the potato, which combines with the starch of the tuber when cooking at more than 120 °C, such as when cooking French fries. Acrylamide is a well-known carcinogen in animals and a possible carcinogen in humans [10]. Genetic modification can lower the levels of enzymes that promote this deleterious transformation. More recent varieties also incorporate resistance to late blight, the resistance gene coming from wild Andean potatoes. These modified potatoes are marketed under the name *Innate*®, and their regulatory status deserves to be examined (Box 10.2).

Box 10.2: Are *Innate*® Potatoes GMOs?
Developed by Simplot under the brand name *innate technology* in 2011, innate® potatoes are characterized by genetic modifications that give them the specific properties mentioned above. It is not a single cultivar but a set of potato varieties that have been subjected to the same transformation technology. The innate® potato does not contain any genes that are foreign to the potato gene pool (hence its trade name, which emphasizes that all the genes in these potatoes are indeed "innate" potato genes).

The first varieties received approval for cultivation in the United States in 2014 (USDA (*US Department of Agriculture*)) and for marketing for food and feed in 2015 (US FDA (*Food and Drug Administration*)). To be approved, genetically modified products are subject to GMO regulations. The first innate® potatoes were therefore authorized under this procedure and labeled "GMO." however, these potatoes are not transgenic products [11] but cisgenic (resulting from cisgenesis²), and the transformation carried out to *silence* the

(continued)

---

²Cisgenesis: gene transfer by genetic engineering belonging to the same species. These genetic modifications can also be carried out with classical hybridization methods, which take longer.

Box 10.2 (continued)

biosynthesis of four proteins involved in the production of acrylamide uses RNAi (interfering) (cf. Chap. 4). After the adoption of the *secure rule, innate®* potato varieties were submitted to the USDA authorities, who granted them the exemption from GMO regulation [12] (see Chap. 9). The varieties in question are therefore no longer labeled as GMOs.

Animal Applications

Biotechnological applications applied to animals have always been more difficult than those applied to plants and microorganisms, due to the biological and functional complexity of higher animal organisms. Obtaining transgenic animals in the laboratory requires a very high level of technical skill, and the expression of the transferred exogenous genes is also variable due to the biological characteristics of the different animal species, which are not homogeneous. A small number of projects (about 40) have been developed between 1985 and 2015 [4].

Commercial applications in animals, apart from the development of transgenic animals for laboratory research and the anecdotal production of the biofortified milks mentioned above, are unique in that only one transgenic animal has been commercialized to date, the *AquaBounty* transgenic salmon, after 25 years of research. This transgenic salmon produces an increased amount of growth hormone, which causes an accelerated growth of the animal and consequently decreases the final food consumption. After obtaining marketing approval in 2017 in Canada and then in the United States, sales amounted to 4.5 tons in 2019 [13]. Recently, in February 2023, following campaigns by anti-GMO activists, the company *AquaBounty Technologies* announced that it would give up raising transgenic salmon in its Prince Edward Island aquaculture farm and that production would be carried out by the one located in Indiana in the United States, keeping the hatchery in Canada, with the genetically modified salmon eggs being sent for development in Indiana and then in Ohio [14].

Conclusion

Genetic engineering has led to breakthroughs in human and animal health through improvements in vaccines and a wider range of therapeutic options.

Genetically modified laboratory animals have made it possible to better understand the mechanisms of observed pathologies and are now essential tools for researchers. But already in some countries, voices are being raised against GMAs (genetically modified animals). How far will this protest go?

Biofortified or better-preserved foods open avenues to improve the nutritional quality of marketed products or to address a major public health problem. According to the FAO (*Food and Agriculture Organization*), 500,000 of the 5 million children suffering from vitamin A deficiency become blind every year for this reason [15]. Golden rice, after 30 years of research and struggle, can finally be grown and marketed in some countries. Let us hope that it will bring a small answer, in addition to other approaches, to fight against this terrible scourge that is xerophthalmia.

Bibliography

1. Callaway, E. (2020). Labs rush to study coronavirus in transgenic animals – Some are in short supply. *Nature, 579*(7798), 183. https://doi.org/10.1038/d41586-020-00698-x
2. Animals Research Info. (2014). *The rat (GM),* https://www.animalresearch.info/en/design-the-research/animals-of-research/the-rat-gm
3. Houdebine, L. -M. (2000). *Le vrai du faux* (p. 236). éditions Le Pommier.
4. Houdebine, L. -M. (2018). Les nouveaux outils des biotechnologies animales. In C. Regnault-Roger, L. –M. Houdebine, A. Ricroch (dir), *Au- delà des OGM. Science- Innovation-Society* (pp. 67–92). Presses des Mines.
5. Ricroch, A. (2018). Biotechnologies végétales: Applications et perspectives agricoles. In C. Regnault-Roger, L. –M. Houdebine, A. Ricroch (dir), *Au-delà des OGM. Science-Innovation-Société* (pp. 93–114). Presses des Mines.
6. Regnault-Roger, C. (Ed.). (2021). *La culture du tabac en France. Sauvegarder un savoir-faire, promouvoir l'innovation?* (p. 253). Presses des Mines.
7. Medicago. (2021, December 16). *Medicago soumet les résultats de la Phase III de son candidat-vaccin contre la COVID-19 produit sur plantes à Santé Canada,* https://medicago.com/en/communique/medicago-submits-phase-iii-results-of-its-plant-based-covid-19-vaccine-candidate-to-health-canada/
8. C-Sciences CS. (2022). *COVID-19: le vaccin Medicago rejeté par l'OMS,* https://www.cscience.ca/2022/03/25/medicago-le-vaccin-tant-attendu-pour-le-printemps-2022/
9. Franche, C. (2018). Biotechnologies végétales et pays en développement. In C. Regnault-Roger, L. –M. Houdebine, & A. Ricroch (dir), *Au-delà des OGM. Science- Innovation-Société* (pp. 115–135). Presses des Mines.
10. ANSES. (2017). *L'acrylamide dans les aliments,* https://www.anses.fr/en/content/l%E2%80%99acrylamide-dans-les-aliments
11. Katiraee, L. (2015, May 27). Scientist mom evaluates Simplot's GMO Innate potato. *Genetic Literacy Project,* https://geneticliteracyproject.org/2015/05/27/scientist-and-mom-evaluates-simplots-gmo-innate-potato/
12. USDA-APHIS. (2021). *USDA announces deregulation extension of potato developed using genetic engineering,* Publ, https://www.aphis.usda.gov/aphis/newsroom/stakeholder-info/stakeholder-messages/biotechnology-news/ge-potato
13. ISAAA. (2021). Global status of commercialized biotech/GM crops in 2019: Biotech crops drive socio-economic development and sustainable environment in the new frontier. ISAAA *Brief No. 55.* ISAAA: Ithaca, p. 100.
14. Lecacheur, J. (2023, February 8). *AquaBounty ne produira plus de saumon génétiquement modifié au Canada,* Radio-Canada, https://ici.radio-canada.ca/nouvelle/1954454/aquabounty-saumon-genetique-ogm-ile-du-prince-edouard
15. Pro visu Foundation. (2020). *Carence en vitamine A et xérophtalmie,* https://www.provisu.ch/en/most-common-diseases/vitamin-deficiency-and-xerophthalmia.html

Chapter 11
Plant GMOs: Agricultural Applications

Varietal improvement of cultivated plants is one of the pillars of agriculture, the objective being to feed the world's population better and better with increased yields, food quality and safety, and more recently the desire to produce not only more and better but also while respecting the environment.

It is therefore not surprising that biotechnological tools were used very early on in this search for progress to develop new varieties with genome modifications using various techniques, including transgenesis in 1983.

What is the purpose of these genomic modifications? For what applications? What has happened to these innovations? Here are a few points of reference.

First Transgenic Plants

The first transgenic plants were the result of a competition between three research teams that presented their results simultaneously at the 15th Winter Symposium in Miami in January 1983 [1]. Two teams had used tobacco as a model, that of Mary-Dell Chilton (who received the nickname *Queen of Agrobacterium*) and that of the Belgian laboratory of the University of Ghent led by Marc Van Montagu who worked with Jeff Schnell (before the latter was appointed director of the *Max Planck Institute* in Cologne), while the third team, that of Robert Fraley of the Monsanto Company, had focused on petunia [2]. The first transgenic plants were born! The tobacco transformed by the Belgian team integrated a gene from the *Bacillus thuringiensis* which gave the plant insecticidal properties.

This was followed in the 1990s by tobacco and tomato plants in China that were genetically modified to resist viruses [3, 4] and in the United States in 1994 by a tomato with delayed ripening, the *Flavr Savr* tomato mentioned in Chap. 2.

C. Regnault-Roger, *Biotech Challenges*, https://doi.org/10.1007/978-3-031-38237-6_11

Produced by the Calgene Company, it proved to be disappointing both for its taste qualities, which were not very convincing, and for the advantages in terms of transport that it should have produced, but which were not observed. The acquisition of Calgene by Monsanto in 1996 put an end to this experiment.

It was at the same time, in 1995, that the first marketing authorizations were granted in the United States and Canada for *Bollgard® Cotton*, carrying the MON 531 transformation event from Monsanto, and for *Knockout® Corn*, carrying the Bt 176 event from Syngenta, both of which are resistant to insects of the lepidopteran family, and then for *New Leaf Potato® Russet Burbank* from Monsanto that is resistant to a Colorado beetle [5]. This was the prelude to the epic of GMO cultivation, which took root mainly on the American continent and in part of Asia (see Chap. 12).

Characteristics of Transgenic Crop Plants

The boom in genetically modified plants (biotech plants) really began in 1996 with the cultivation of transgenic plants marketed to facilitate weed control and to better resist insect pests of these crops.

Indeed, the first transgenic plants are modified to preserve the health of the plant against its bio-aggressors by integrating resistance genes against aerial or underground insects or nematodes (worms that damage the roots of the plant), major pests that attack a plant in the process of croping, or against pathogens that spread diseases.

The first transformation events introduced into these genetically modified plants consist of genes that code for proteins toxic to insect pests that are decimated. These plants are subjected to fewer attacks and are more resistant to aggression (hereafter referred to as IR "insect-resistant").

To control weeds that compete with crops for water and soil nutrients in the field, another approach is used: it consists of introducing a herbicide tolerance gene into the plant. This gene produces an enzyme that allows it to detoxify the herbicide. As a result, the plant resists the destructive activity of the phytosanitary product. If a nonselective herbicide (also called "total" because of its broad spectrum of activity) is chosen, such as glyphosate or glufosinate, only the plants that possess the detoxification enzyme are resistant to the herbicide and are not destroyed, while other plants, which are undesirable in the field, are. These undesirable plants are regrowths of previous crops and also weeds, such as bedstraw and poppies which reduce yields, as well as weeds such as the deleterious *Ambrosia* or *Datura* which secrete natural molecules that are toxic to humans and animals. Only the transgenic plants tolerating the herbicide, which is fatal for other plants, survive. These plants will later be called HT: "herbicide-tolerant."

The dominant transformation favored by farmers is tolerance to a herbicide (HT). This innovation, which makes it easier for them to clear their fields of undesirable plants, is rapidly surpassing the area cultivated with IR crops. Other varieties have incorporated drought resistance to cope with climate change in some regions. Since 2007, trait stacking (i.e., the fact that in the same plant there can be several transformation genes) has become more important: these plants better meet both agronomic and climatic requirements. These versatile plants are preferred to those with only one transgenic trait. As a result, HT crops are down 10% since 2014, and those resistant to one or more insects (IR) are stagnating around 10% of the biotech acreage grown. Thus, in 2019, 45% of biotech crops carry several stacked genetic modifications: tolerance to one or more herbicides (glyphosate, glufosinate, dicamba, 2-4D, isofluxatole, etc.), insect resistance (IR), but also drought tolerance. In the United States, the progress is remarkable: the area cultivated with transgenic plants tolerant to water deficit has increased 40-fold in 5 years. In Africa, the WEMA (*Water Efficient Maize for Africa*) project is developing a maize adapted to drought (see Chap. 12).

There is also a broadening of the range of genetic modifications to improve the food quality of commercialized GM products. Examples include *Plenish* soybeans, whose oil contains more unsaturated oleic fatty acids and less saturated fatty acids than conventional soybeans, thus improving their nutritional profile, or cottonseed cake for animal without gossypol, a polyphenol that is toxic to some species.

Which Transgenic Plants Are Grown?

Four major transgenic crops are now grown on a large scale: soybeans (48.2% of the world's cultivated area), maize (32%), cotton (13.5%), and canola (Canadian spring rape) for 5.3%. The proportions of adoption of biotech varieties compared to conventional varieties remain stable: three quarters or more of the world's cotton and soybean crop is transgenic and nearly one third for maize and canola. In the United States, they reach 98% for cotton (up 4% since 2018), 92% for corn, and 94% for soybeans.

More regional crops account for 1% of the total biotech acreage cultivated: rice, glyphosate-tolerant alfalfa (*Roundup Ready*®), papaya and squash, potato or sugar beet or pineapple (in Costa Rica), and ornamental flowers (Colombia). A success on a small and localized crop has been noted in recent years: in Hawaii the planting of transgenic papaya resistant to a disease caused by a virus *Papaya ringspot virus* (PRSV). Hopes are also pinned on the development of genetically modified plum trees to combat sharka, the most devastating viral disease of the *Prunus* genus caused by *plum pox virus* (PPV). Originally identified in the Balkans, this pathology has spread widely, particularly in North America.

In local success stories, there is also a high adoption of brinjal Bt (RI) eggplant by farmers in Bangladesh (Box 11.1).

Box 11.1: Brinjal Bt Eggplant: A key Crop in Bangladesh

Eggplant (brinjal, *Solanum melongena*), native to India, is an economically important crop for the countries of Southeast Asia. It provides the poor populations of the Indian subcontinent and the Philippines with important food supplements in essential nutrients (e.g., vitamins). It is ravaged locally by a lepidopteran (*Leucinodes orbonalis*), the eggplant fruit and shoot borer, which causes heavy losses, up to 86% of the harvest in Bangladesh. Repeated and close insecticide treatments (between 20 and 40 treatments) are necessary, which is not without consequences for the environment and for the health of the farmers. The applications are indeed carried out with ancient and rudimentary means (backpack sprayers carried without personal protective clothing).

Extensive research has been conducted in India to find a solution to the ravages of the insect. It was carried out by Mahyco (*Maharashtra hybrid seeds company*), the first seed producer in India, in collaboration with the University of Tamil Nadu, in order to obtain local varieties of Bt transgenic eggplants resistant to the lepidopteran pest.

Under pressure from powerful anti-GMO NGOs and despite conclusive experimental field trials, in 2010, the Indian authorities, followed by those of the Philippines in 2011, decided on a moratorium.

In response, the government of Bangladesh, building on the work of national research organizations *Bangladesh Agricultural Research Institute (BARI), Bangladesh agricultural development corporation (BADC)*, and *Department of Agricultural Extension (DAE)*, decided to give Bt eggplant seeds to farmers for field trials.

This crop mobilizes in fact in this country 150,000 farmers on 50,000 ha (these are very small farms). Initially, 20 farmers started growing brinjal Bt eggplant on 2 hectares in 2014. The success was meteoric: In 2020, 27,000 farmers are growing it. This rapid adoption is the result of telling numbers: a 45% decrease in lepidopteran borer infestations, a 61% pesticide savings, and a six-fold increase in income per hectare from the crop [5]. Economic impact studies in five districts of Bangladesh are underway [6].

What Is the Need for These Biotech Crops?

In the societal controversy raging over GMOs, a lot of false information and preconceived ideas are circulating to demonstrate the harmfulness of the cultivation of genetically modified plants, a cultivation that has been practiced for 25 years now: a quarter of a century that has demonstrated that the assertions the Cassandras are spreading are not proven, as was moreover emphasized in the report of the three American academies published in 2016 already mentioned in Chap. 7.

Here are some elements that highlight the benefits of cultivated GMOs [7].

GMOs to Respond to Climate Change

Faced with this great challenge of preparing for the future with a changing climate, GMOs are proving to be answers that can intervene at different levels by limiting agricultural CO_2 emissions and by dealing with the emerging risks of new pests in regions where they did not exist before.

Reasonable Use of Phytopharmaceutical Inputs and Less CO_2

By giving the biotech plant new properties to secrete, for example, insecticidal substances to resist the pests that affect its crop, it is less necessary to treat with phytopharmaceutical products, whether synthetic or of biological origin (biocontrol products). Crop management is modified, and phytosanitary technical itineraries are improved with a reduction in preventive or curative treatments.

Indeed, the farmer is, each year, confronted with the climatic conditions of the moment on the crop. Even if he has reduced the risks by taking care of the state of his soils, by carefully choosing the varieties he will plant in a pattern of crop rotations and controlled ecosystems, and by consulting the profession's information bulletins that inform him of foreseeable pest events, he is faced with the dilemma of prevention or cure. Prevention in crop management is in fact a speculation on a future that does not always happen; in some cases, we have won by preventing an infestation from developing, but in others, we have treated for nothing because the weather has changed and the conditions for the installation of the parasite are no longer met. And that's where a farmer's know-how lies with the technologies he has at his disposal, to treat at the right time, in the right place, and at the right dose.

By giving them a better ability to limit the effects of their bio-aggressors, one of the consequences of the genetic improvement of cultivated plants is to allow the adjustment of insecticide, fungicide, and herbicide treatments for crop management as well as to reduce comfort treatments, i.e., the administration of phytosanitary products on a preventive basis. These products, which aim to preserve or treat plants, must be used within the framework of a reasoned practice.

The resulting savings in plant protection products reduces the risk of pesticide residues in food and improves food safety. It also has environmental benefits by reducing the risk of pesticide contamination of the biosphere and greenhouse gas emissions from the use of farm machinery in the field. It also has economic benefits by reducing the cost of inputs and fuels as well as the working time of the farm operator.

These environmental benefits apply to all biotech crops. Two experts in biotech economics, Graham Brookes and Peter Barfoot, have been tracking the environmental impact of biotech crops annually since 1996. In their latest available study, conducted over a 22-year period, they calculated that biotech crops can be credited in 2018 with a reduction in CO_2 emissions of 23,027 tons. This saving is linked to changes in agricultural practices with simplified cultivation techniques (SCTs) and reduced spraying of plant protection products, which also reduce the amount of fuel consumed by agricultural machinery. SCTs also make soils less vulnerable to erosion and increase soil biodiversity (earthworms, soil insects such as springtails). The reduction of active substances consumed during insecticide and herbicide treatments was calculated. Over the 22-year period mentioned above, it is estimated at 775,400 tons. In addition to this evaluation, an indicator, the *Environmental Impact Quotient* (EIQ), was established for each active plant protection substance in the field. The authors note an improvement in this quotient of around 18.5%, as well as an 8.3% decrease in the quantities of pesticides used [8]. Plant protection technical itineraries have improved and have become more virtuous.

Dealing with Emerging Risks

Among the emerging risks, the risks of fusariosis and mycotoxins have increased in Europe for corn and wheat crops due to climate change. Indeed, due to the very hot summer, the installation of new tropical pathogenic fungi such as *Aspergillus* secreting aflatoxins is added due to the classical toxigenic fungi of the *Fusarium* genus. Climate change can be accompanied, when the conditions are right, by an increase in the content of mycotoxins, some of which were previously unusual.

Thus, in recent years, several contaminations by mycotoxins of marketed food products have been noted. They could have had dramatic consequences without the vigilance of Health Control Services. In 2013, milk imported from Eastern Europe was refused in France because of high levels of mycotoxins: the cows had consumed fodder contaminated by mold. In May 2014, this time concerning supermarkets, it was Auchan *Organic Corn Cakes* and *Bio-village Wheat Bran* from the *Repère* brand (Leclerc) that were hastily withdrawn from shelves because of levels of a mycotoxin (DON (deoxynivalenol)) above regulatory thresholds. In June 2014, the Italian newspaper *Il Fatto Emilia Romagna* reported that wheels of Parmesan cheese were contaminated with aflatoxins. This contamination occurred during the heat of the August 2013 when cows were fed grain corn or corn silage containing *Aspergillus* that had developed. The health authorities of the various European countries noted that under favorable climatic conditions, these *Aspergillus*, which are endemic in Africa and contaminate poorly preserved cereal crops, can also proliferate in the South or East of Europe.

The risk of mycotoxins is well known in Europe since chronicles of the Middle Ages echoed the *Mal des Ardents* or *Feu de Saint Antoine* (*Saint Anthony's fire*) which provoked hallucinations and circulatory disorders that could lead to the death of poor people who consumed moldy food. An intoxication was described in the 1950s with the consumption of bread in Pont-Saint-Esprit which resulted in the death of 10 people and about 50 psychiatric internments.

Various pathologies can indeed occur in mammals following the consumption of mycotoxins: renal, liver, digestive tract, immune system, and reproductive system disorders. The nature of the symptoms depends on the mycotoxin that contaminates the food. Some mycotoxins are carcinogenic or immunotoxic; others are endocrine disruptors. Severe diseases have been observed in humans following the consumption of contaminated flour (e.g.,Pont-Saint-Esprit), and animal feed is not left out. Historically, the first case of British turkey farms being decimated by aflatoxin-contaminated peanuts imported from Senegal was described in 1960 [9]. Horses were poisoned by fodder or bedding contaminated with mycotoxins. The awareness of the mycotoxin risk led the European health authorities to establish a regulation in 2007 setting thresholds not to be exceeded.

Research has shown that GMO maize MON 810 not only reduces the number of insect pests that consume the stalks or ears because they are decimated by the Bt toxin but also reduces the levels of mycotoxins in the crops [10]. This corn is genetically modified by incorporating a gene to allow the plant cells to synthesize the Cry1Ab protein that disrupts the digestive function of two major pests of this crop, the European corn borer (*Ostrinia nubilalis*) and the pink stem borer (*Sesamia nonagrioides*), and kills the insects. The reduction of phytophagous insects leads to a reduction in the number of wounds that they inflict on the plant while feeding and through which pathogenic fungi, agents of fusariosis and producers of mycotoxins, settle.

By reducing the incidence of fusariosis and the risk of mycotoxins, the cultivation of MON 810 maize improves the sanitary quality of crops. An Italian study showed that piglets fed with Bt MON 810 transgenic maize had normal growth, while those fed with conventional maize contaminated with mycotoxins showed significant weight loss [11].

Thus, the cultivation of this transgenic corn allows to significantly reduce the mycotoxin risk in the harvests. It is a response to the increasingly hot summers due to global warming. This is why it is very much appreciated by Spanish farmers who face high temperatures in summer. And the ban on growing MON 810 corn in France by law 2014–567 is not understandable, from a scientific and public health point of view. Indeed, the European corn borers (*Ostrinia nubilalis*) and pink stem borers (*Sesamia nonagrioïdes*) that ravage corn crops are increasingly present in France in areas where they did not previously exist. Their progression toward more northern regions continues. The fall 2018 *Bulletin de Santé du végétal* de la Région Centre-Val de Loire described instances of two annual generations of European corn borer instead of one in Champagne Beauceronne, Sologne-Val de Loire, and Touraine. The previously absent Sesamia has been captured regularly there since 2010, and its range has extended north of the Loire River since 2016 [12]. The close renewal of these insect pests then requires several treatments during the season (and nothing like this with Bt MON 810 maize since the plant protects itself).

Improving Agricultural Biodiversity

Contrary to popular belief, the cultivation of biotech plants promotes agricultural biodiversity. Indeed, the commercialization of GM plants has led to an increase in the number of varieties cultivated. Bernard le Buanec, a member of the French

Academy of Technology and of the French Academy of Agriculture, gives some figures on this subject: in 2016, there were 87 different varieties of Bt corn including the MON 810 transformation event grown on 35,000 ha in Catalonia and in India 205 varieties of cotton grown including the MON 531 transformation event as well as 309 varieties for the MON 15965 trait [12]. Jean-Claude Pernollet mentions more than 4300 varieties of transgenic corn marketed under 202 different brands and produced by 173 companies [13].

The cultivation of transgenic plants that are resistant to ubiquitous diseases contributes to saving biodiversity in the territories by preserving species that are threatened by the pathogens they are associated with. Two local crops that we mentioned earlier (page 83), plum trees saved from sharka (*plum pox virus* (PPV)), and also the *papaya ringspot virus* (PRSV) crop in Hawaii, became emblematic because their cultures were strongly contested by protest from anti-GMO activist movements. The American film *Food Evolution*, whose screening in French cinemas was blocked by anti-GMO activists, illustrates this fight to save the papaya trees in Hawaii.

The cultivation of GMO plants is also positive for the preservation of entomological biodiversity on the plots. Auxiliary insects, which are beneficial to crops, are more numerous on plots cultivated with a transgenic plant: whereas the application of an insecticide kills most of the insects in the field, the biotech plant specifically kills only those that attack it by poisoning them. An increase in biodiversity in GM cotton plots has been observed in China [14]. While the damage of the cotton moth was reduced by the toxicity of the plant to this insect, the reduction of insecticides applied increased the populations of three major groups of beneficial insects: ladybugs, spiders, and lacewings. A similar watching was made by Claude Ménara, a farmer in southwestern France (Lot-et-Garonne) when he was growing MON 810 maize on his farm, i.e., before the 2014 ban [15].

Respect Good Agricultural Practices to Prevent Foreseeable Risks

Whenever crop pests are controlled whether by conventional methods using chemical (mineral or synthetic) or biological phytopharmaceutical formulations, resistance phenomena are to be expected according to Darwinian patterns of coevolution.

For example, insecticide formulations based on *Bacillus thuringiensis* (Bt bacteria), which are widely and repeatedly used in organic agriculture, have provoked numerous resistances of the European corn borer in the field against which they were used. Similarly, the repeated use of the same herbicide on the same plot where the same crop is grown year after year, leads to the appearance of weeds resistant to this herbicide.

The use of biotech plants to control insect pests or to help the farmer to better weed his field does not free him from observing good agricultural practices.

In order to delay these predictable resistances, it is recommended in the case of transgenic crops to set up "refuge zones" cultivated with conventional plants adjacent to the plots cultivated with transgenic plants, so as to operate a genetic mixing between the insects feeding in the different zones, and to reduce or delay the appearance of resistances in the target insect. These refuge zones were set at 20% of the area of the plot cultivated with a biotech plant incorporating a single transformation event.

Several patterns were recommended: alternating strips in the field or encircling or side by side (Fig. 11.1). A maximum distance between transgenic and conventional plants should not exceed 750 m. The cultivation of biotech plants incorporating several transformation elements, by multiplying insecticidal toxins, is also a response to delay the appearance of resistance. The percentage devoted to refuge areas is fixed according to the crop. For example, in cotton production areas, the *US Environmental Protection Agency* (EPA) has set a percentage of refuge zones of 50% and for GM maize crops containing several Bt genes (stacking) a refuge zone of 5%.

In order to reduce the workload of farmers, a new process is proposed called *refuge in the bag* (*RIB technology*): the seed bag contains a specific mixture of Bt and non-Bt seeds set according to the required percentage of refuge areas for the

Fig. 11.1 Refuge areas in transgenic plant plots. (Diagram courtesy of Ogm.Gouv.qc.ca, photo credits Catherine Regnault-Roger)

transgenic variety grown. The farmer sows both types of seeds in one pass and then scatters them in his field. This technology has the advantage of encouraging and obliging the farmer to observe the establishment of refuge areas in his field, which is not always the case when the field has to be structured into distinct zones. Because the exercise can be very complex with Bt corn containing several gene stacks, such as the *Genuity® SmartStax™* Bt corn varieties, which contain six different genes (*cry1A.105, cry2Ab2, cry1F,* and *cry3Bb1, cry34Ab1,* and *cry35Ab1*) to protect them from aerial (lepidopteran) and subterranean (beetle) pests. For this reason, the US EPA approved the RIB technology in 2010 based on successful mathematical models [16].

To prevent resistance to a herbicide, it is also necessary to observe good agricultural practices (GAP) by rejecting the practice of a monoculture or the repeated use of the same herbicide from one crop to another, which inevitably leads to the acceleration of the development of resistance.

The case of glyphosate is eloquent in this respect. Because it is a nonselective herbicide among the least toxic and most effective, it was already used extensively long before the marketing of biotech plants. The first case of resistance to this herbicide in a weed, ryegrass (*Lolium rigidum*), was described in Australia as early as 1996 in a conventional farm before cultivation of transgenic crops.

The widespread use of glyphosate-tolerant biotech plants (soybean and corn in particular) and the success of this easy-to-use herbicide, which is very active at lower doses, among farmers, led to the emergence of resistance in several weeds. The use from 1 year to the next of *Roundup Ready®* corn (glyphosate-tolerant HT corn) grown as a monoculture or of rotations of this crop alternating with *Roundup Ready®* soybeans (HT soybeans tolerant to the same herbicide), as is done in certain regions of the North American continent that we visited, leads to an increase in this risk of resistance. It is necessary to diversify the active substances in plant protection treatments. This is why in recent years, new varieties that are multi-tolerant to several herbicides (glyphosate, dicamba, glufosinate, 2-4D, isofluxatole) have been registered and have arrived on the North American market, which allows the farmer to diversify his phytopharmaceutical itineraries: because growing biotech plants does not dispense with respect for agronomic rules! But it would also be necessary for seed stores and distributors in very rural areas of North America to give operators a choice by broadening the range of products they sell! Only through the cooperation of all players in the industry can this phenomenon be limited. It is a commonsense approach to preserve a precious tool for sustainable agriculture!

Is the Coexistence of Biotech and Organic Farming Possible?

The coexistence of the different agricultural practices such as conventional agriculture, conservation agriculture, biotech or precision agriculture, had never been a concern before organic agriculture was developed and is strongly encouraged in recent years by the French and European public authorities, which have granted very substantial specific subsidies. This type of agriculture has the particularity of having an obligation of means and not of results. Indeed, its followers must observe a very precise and restrictive specification on the methods to be followed for the conduct of the crops. For ideological reasons, the cultivation of biotech plants and the use of certain synthetic organic pesticides are refused, while ecotoxic mineral chemical pesticides such as copper-based sprays or hi-tech varieties such as Renan wheat are accepted (see Chap. 6). Let us not look for logic! Because of these resource constraints, a detected adventitious presence of GM plants above 0.9% in crops grown on organic plots would have a negative economic impact on the value of the crops, which would no longer be considered "organic" (organic products are generally sold at a higher price than those from conventional agriculture [17]).

The unintentional dissemination of transgenes between different species is a natural phenomenon (see Chap. 3). The gene flows has to be considered differently depending on whether the considered species are interfertile or cannot hybridize with each other, according to Jean-Claude Pernollet in his 2018 review of the environmental risks and benefits of biotech plants [18]. In the second case, he points out, "the insertion of a viable transgene and its selection require a lot of time so that it is not observable on a human time scale," while, in the first case, measures for the coexistence of cultivated varieties will have to be taken by observing distances to be respected between plots or by staggering sowing, if we absolutely want to avoid incidental hybridizations which, let us remember, have no negative health and environmental consequences. Is organic agriculture justified in imposing its production scheme on other agricultures? A scheme that appears to be based on ideological *preconceptions* that are not scientifically founded.

To provide a pragmatic response to this concern, a reflection was conducted by the Portuguese authorities to organize the coexistence between biotech (GM), conventional, and organic agriculture. In Portugal, this coexistence is guaranteed by Decree-Law 160/2005, which established administrative rules and technical standards. These must be based on scientific and technical progress (Article 8) and have led to the establishment of dedicated geographical areas. The regions of Lisbon and Alentejo have land use characteristics (size and distribution of farms in particular) that are favorable to the coexistence of different types of agriculture (biotech, conventional, and organic). In these two provinces, Portuguese biotech plants have been cultivated for more than 15 years and coexist with organic crops.

The European Union has also taken up the subject with a European research program called PRICE (*PRactical Implementation of Coexistence in Europe*). It delivered its conclusions in 2015, and demonstrated through experiments in Spain that coexistence between GM and non-GM crops is possible. Measures such as the use of plants that are unable to reproduce by self-pollination (sterile males), staggered sowing, or the creation of buffer zones should be applied. The application of these technical measures in the experiments has ensured that the crops comply with the official European labeling threshold for adventitious presence of GMOs of 0.9%.

Conclusion

It appears from this overview that biotech plants are proving to be very useful in various ways.

They provide answers to the problems posed by crop pests. These not only cause economic losses for the producer by destroying part of the harvest or by making it unfit for sale, but they also cause health problems. The example of mycotoxins produced by pathogenic microscopic fungi, which settle in favor of climatic conditions favorable to the proliferation of harmful insects in the fields, is edifying. Food intended for humans (cereals, milk, cheese) are contaminated in an insidious and invisible way and can cause serious intoxications if the sanitary quality of food is not controlled. Animals that eat contaminated crops with levels just below the thresholds also become chronically ill.

Biotech plants also provide answers to climate change with, on the one hand, varieties adapted to drought and, on the other hand, a reduction in greenhouse gas emissions by agricultural machinery.

They reduce the arduousness of the farmer's work by improving the technical itineraries on the farms. The reduction in the number of phytosanitary treatments results not only in substantial savings in the purchase of phytopharmaceutical products but also in better yields, with the direct consequence of improving the health of small farmers in developing countries in tropical zones. Indeed, the use of biotech plants avoids the rustic conditions of phytosanitary product spraying. These are carried out very frequently in hot countries under rough conditions and are often accompanied by the most basic personal protective equipment. This is one key of success mentioned by the small farmers in Bangladesh who have adopted the brinjal GM Bt eggplant crop.

Biotech plants are thus important tools to be taken into consideration in order to protect human, animal, plant, and environmental health, which, as we now know, is part of the global approach of *One Health* in the world.

Bibliography

1. NIHF-The Smithsonian National Museum of American History. (2022). *Mary-Dell Chilton Transgenic Plant,* https://www.invent.org/inductees/mary-dell-chilton
2. World Food Prize Foundation. (2022). *2013: Van Montagu, Chilton, Fraley,* https://www.worldfoodprize.org/en/laureates/20102019_laureates/2013_van_montagu_chilton_fraley/
3. Tao, Z., & Shundong, Z. (2003). GMO use in China: Issues and debates. *Chinese Perspectives, 76,* 52–60.
4. James, C. (1997). Global status of transgenic crops in 1997 ISAAA *Briefs No. 5.* ISAAA (Ithaca, NY), p. 31, https://www.isaaa.org/resources/publications/briefs/05/default.html.
5. Shelton, A. M., Zhao, J. Z., & Roush, R. T. (2002). Economic, ecological, food safety and social consequences of the deployment of Bt transgenics. *Annual Review of Entomology, 47,* 845–881.

6. Shelton, A. M., Hossain, M. J., Paranjape, V., Azad, A. K., Rahman, M. L., Khan, A. S. M. M. R., Prodhan, M. Z. H., Rashid, M. A., Majumder, R., Hossain, M. A., Hussain, S. S., Huesing, J. E., & McCandless, L. (2018). Bt eggplant project in Bangladesh: History, present status, and future direction. *Frontiers in Bioengineering and Biotechnology, 6*, 106. https://doi.org/10.3389/fbioe.2018.00106

7. Shelton, A. M., Sarwer, S. H., Hossain, M. J., Brookes, G., & Paranjape, V. (2020). Impact of Bt Brinjal cultivation in the market value chain in five districts of Bangladesh. *Frontiers in Bioengineering and Biotechnology, 8*, 498. https://doi.org/10.3389/fbioe.2020.00498L

8. Regnault-Roger, C. (2020). *Des plantes biotech au service de la santé végétale et de l'environnement* (p. 56). Fondation pour l'innovation politique.

9. Brookes, G., & Barfoot, P. (2020). Farm income and production impacts of using GM crop technology 1996-2018. *GM Crops & Food, 11*(4), 215–241.

10. Regnault-Roger, C. (2017). Comment prévenir le risque de mycotoxines dans la production agricole? *Science and Pseudoscience, 322*, 45–48.

11. Regnault-Roger, C., & Delos, M. (2011). L'intérêt des plantes génétiquement modifiées pour la production agricole: le cas du maïs Bt. In A. Ricroch, Y. Dattée, & M. Fellous (dir.), *Biotechnologies végétales* (pp. 200–211). Vuibert.

12. Rossi, F., Morlacchini, M., Fusconi, G., Pietri, A., & Piva, G. (2011). Effects of insertion of Bt gene in corn and different fumonisin content on growth performance of weaned piglets. *Italian Journal of Animal Science, 10*, 2. https://doi.org/10.4081/ijas.2011.e19

13. Le Buanec, B. (2018). La diversité génétique en agriculture. In C. Regnault-Roger (dir.), *Idées reçues et agriculture* (pp. 169–187). Presses des mines.

14. Pernollet, J. –C. (2018). Plantes génétiquement modifiées. In C. Regnault-Roger (Ed.), *Idées reçues et agriculture* (pp. 188–204). Presses des mines.

15. De Lacour, G. (2012, June 12). Le coton GM favorise la biodiversité, selon une étude chinoise. *Journal de l'environnement*, http://www.journaldelenvironnement.net/article/gm-cotton-favors-biodiversity-according-to-a-Chinese-study

16. Menara, C. (2007, June 4). Mon maïs transgénique est le plus écologique! *Figaro actualités*, http://premium.lefigaro.fr/actualite/2007/04/06/0100120070406ARTFIG90080claude_menara_mon_mais_transgenique_est_le_plus_ecologique.php

17. U.S. Environmental Protection Agency. (2021). *Insect resistance management for Bt plant-incorporated protectants,* https://www.epa.gov/regulation-biotechnology-under-tsca-and-fifra/insect-resistance-management-bt-plant-incorporated

18. Kressman, G. (2021). *Quel avenir pour l'agriculture et l'alimentation bio?* (p. 56). Fondation pour l'innovation politique.

Chapter 12
Cultivated GMOs: What Geopolitical View in 2022?

It is now 25 years since the first transgenic crops were commercialized, and the world's cultivated area is now around to 190 million hectares. This adoption of the innovation in some countries or its rejection was noticed very early. Between 1996 and 2018, the area under biotech crops has grown steadily, but in 2019, a slight decrease of 1.3 million ha is noted. For the first time in 25 years, the increase in acreage is not at the "annual rendezvous"! Is this temporary or cyclical? How should it be interpreted? What is the situation in 2022 in the different parts of the world?

We thought it would be interesting to give some pointers on the prospects for biotech plants at a time when agricultural applications of NGTs will take off.

It was indeed possible to precisely follow the evolution of the cultivation of GM plants since the very beginning thanks to the determination of Dr. Clive James who created in 1991 an international nongovernmental and nonprofit organization, ISAAA (*International Service for the Acquisition of Agri-biotech Applications*). The objective was to develop a humanitarian action of knowledge sharing on bio-technologies and transfer to developing countries through public/private partnerships. Clive James[1] was supported in his approach by American agronomist Norman Borlaug, winner of the 1970 Nobel Peace Prize for his humanitarian action in favor of food and the "Green Revolution" in developing countries.

Each year since 1996, this organization, hosted by Cornell University since 1992, has published an authoritative annual report. It shows the world's acreage of biotech-modified crops and highlights the development of these crops on five continents. The 24 reports, published and made available online free of charge until 2019 (when Clive James, now emeritus, left the organization), are a valuable source of information on GMO crops. In January 2022, the ISAAA organization announced that its

---

[1] Dr. Clive James, a graduate of *Aberystwyth University* (MSc) and PhD of Cambridge University, was honored in 2011 by the Welsh University for the same reason as the Nobel Prize (a career focused on humanitarian action and developing countries), https://www.aber.ac.uk/en/news/archive/2011/07/title-101,413-en.html

© The Author(s), under exclusive license to Springer Nature Switzerland AG 2023
C. Regnault-Roger, *Biotech Challenges*,
https://doi.org/10.1007/978-3-031-38237-6_12

work would be incorporated into the *BioTrust Consortium*. The page of Dr. Clive James' tremendous work in identifying the global rise of biotech crops is now turned out!

First-Generation Agricultural GMOs: Rapid Expansion

In 1996, the cultivation and commercialization of biotech plants really began, and their growth can be measured by the expansion of the cultivated areas ever since.

But before that, between 1984 and 1995, a great many field experiments (25,000) had preceded the cultivation and marketing of biotech plants. These field trials had been carried out in 45 countries, but nearly three quarters of them were in North America, in the United States, and in Canada. The other 7000 trials were in Europe, South America, and Asia, as well as a few trials in South Africa [1].

Authorization procedures for these new varieties were established (see Chap. 6), and five countries embarked on the adventure in 1996, the United States, Canada, Argentina, Australia, and Mexico, along with a sixth country, China, a pioneer country that had already commercialized transgenic tobacco and tomatoes in the early 1990s.

In 1996, the area of transgenic plants covered 1.7 million ha, of which 51% were in the United States and 39% in China while Canada and Argentina each cultivated 4% and Australia and Mexico less than 1%.

In 1997, 11 million ha of transgenic plants were cultivated, more than six times the area of the previous year.

Ten years later, in 2007, the area under biotech crops increased by a factor of ten, with 114.3 million ha, a jump of 105.1 million ha. The increase continues, peaking in 2018 at 191.7 million ha, representing an increase of more than 77 million ha over the past decade. Table 12.1 and Fig. 12.2 show the situation in 2019.

What Crops Are Being Developed?

In 1996, the first year of commercialization of biotech plants, the species cultivated were tobacco (35% or one million ha), cotton (27% or 800,000 ha), soybeans (18% or 500,000 ha), corn/maize (10%), canola (5%), tomatoes (4%), and potatoes (less than 1%). The transformation events concern virus resistance (40%) and insect resistance (37%) as well as herbicide tolerance (23%).

By the end of 1997, the cultivated area had multiplied considerably (11 million ha). The pattern of four dominant crops (soybean, maize, cotton, rapeseed/canola) that persists to this day was established, as well as the hierarchy of transformation elements: herbicide tolerance (HT) followed by insect resistance (IR). That year, herbicide-tolerant plants represented 63% of the cultivated area, insect-resistant

Table 12.1 Main countries producing biotech plants Cultivated area by country from 2006 to 2019 (in millions of hectares) (author's research based on ISAAA data)

Country rankings base year 2019		2006	2010	2014	2018	2019
1	United States	54.6	66.8	73.1	75	71.5
2	Brazil	11.5	25.4	42.2	51.3	52.8
3	Argentina	18	22.9	24.3	23.9	24.0
4	Canada	6.1	8.9	11.6	12.7	12.5
5	India	3.8	9.4	11.6	11.6	11.9
6	Paraguay	2.0	2.6	3.9	3.8	4.1
7	Pakistan	–	2.4	2.9	2.8	2.5
8	China	3.5	3.5	3.9	2.9	3.2
9	South Africa	1.4	2.2	2.7	2.7	2.7
10	Bolivia	–	0.9	1.0	1.3	1.4
11	Uruguay	0.4	1.1	1.6	1.3	1.2
12	Philippines	0.2	0.5	0.6	0.6	0.9
Total world surface		**102**	**148**	**181.5**	**191.7**	**190.4**

plants 30%, and virus-resistant plants 7%. Soybeans, corn, cotton, and rapeseed (canola) accounted for 86% of the GM crops, and 75% of the acreage was in North America.

Twenty-five years later, the area of land cultivated with biotech soybeans compared to conventional soybeans has become the majority (74% or 92.4 million ha in 2019), for cotton too (79% in 2019 or 25.6 million ha). Today, more than three quarters of the soybeans and cotton harvested come from GMO plants.

The area of biotech corn has also increased, from 24% of the total global area of this crop in 2007 to 31% in 2019 or 60 million ha. For canola, the increase is about 10% with 10.2 million ha cultivated in biotech canola in 2019.

IR plants account for 13% and HT plants for 47% of the cultivated area, while 40% of the area is cultivated with multistack varieties (HT/IR) capable of resisting insect pests and at the same time being more easily weeded (gene stacking).

In addition to these four major crops, new biotech plant species are appearing on the market that are grown on small scales, sugarcane, alfalfa, papaya, potato, sugar beet, squash, and eggplant, which represent only 1% of the area cultivated with biotech plants.

A World Divided

Not all countries grow biotech crops. In fact, the world is separated into two subsets as was pointed out (Chap. 9): countries that produce, consume, and export biotech crops and those that refuse to cultivate them but import and consume them after authorization has been granted by national authorities. The first category is located in the North and South American continent, Asia, and Pacific area and the second in the European continent (and its Siberian extension), Africa, the Mediterranean Basin, and the Middle East area. The European Union is divided, with two countries, Spain and Portugal, which cultivate them and a majority which, like France, is opposed to them. The situation is therefore quite contrasted according to the continents. What picture can be drawn?

The American Continent: The Land of Choice for Biotech Plants

The American continent, north and south, accounts for nearly 90% of the land cultivated with biotech plants in 2019.

North America

The United States is undoubtedly the leading country with the largest area from the beginning in 1996 to the present: 75 million ha cultivated (Table 12.1). It is also the country with the greatest diversity of commercialized transgenic crops and the largest number of trials. Corn (45% of the area under corn in 2017 is biotech), soybean (45%), cotton (6%), alfalfa (2%), canola (1%), and the rest are more marginal (sugar beet, papaya, squash, *Innate® potatoes*). Since 1996, over 20 years, the United States has approved 197 processing events in 19 crop species, including 43 for corn, 43 for potato, and 25 for soybean. These crops are produced by 420,000 farmers.

Canada was the first country to allow the commercialization of canola (see Chap. 10) in 1996. Today, it ranks fourth in terms of biotech acreage (12.75 million ha in 2018 and 12.46 million ha in 2019) but is also showing a strong development momentum. The main crops are canola grown on 67% of the biotech area, followed by soybeans (19%) and corn (13.5%). The more marginal biotech crops of sugar beets (15,000 ha in 2018 and 19,000 ha in 2019) and alfalfa (4000 ha in 2018 and 4200 ha in 2019) have increased, while potato (65 ha in 2018 and 40.5 ha in 2019), a crop sold to the United States, has decreased. Since 1996, 177 processing events have been approved by Canadian authorities, including most recently triple stacking (see Chap. 11) for corn and potato.

Latin America

Some Latin American countries such as Mexico and Argentina or Uruguay were among the pioneer countries that hosted trials between 1984 and 1995 and the first field crops in 1996.

Argentina is now the third largest producer of biotech crops with 24 million ha planted in 2019, representing 13% of the world's biotech acreage. Biotech soybean crops account for 73% of the 24 million ha of GM crops grown, followed by biotech corn (up to 5.9 million ha) and the more marginal GM cotton (490,000 ha in 2019, 370,000 ha in 2018). These crops cover more than half (60%) of the country's arable land. The soybeans grown are all GM; the adoption rate of biotech corn is almost 100% (97% in 2017). Most varieties grown are herbicide-tolerant (83% of HT soybeans) or IR/HT multipurpose (83% of corn). Note a strong increase in HT/IR soybeans marketed under the Intacta™ brand, which were grown on 70,000 ha in 2015 and around 4 million ha in 2019 (i.e., 23% of the biotech soybeans grown, demonstrating that it meets a need). New processing events for drought and salinity adaptation, approved in 2015 by Argentine authorities, were developed by the *National Institute of Agricultural Technology* (*Instituto Nacional de Tecnología Agropecuaria*, (INTA Oliveros), of the *National University of Rosario*, and a GM alfalfa with low lignin content (which increases the digestibility of animal feed) received authorization for cultivation in 2018 (third worldwide authorization granted after those of the United States and Canada) and its cultivation started with 1000 ha in 2019 in the Pampa. The transgenic wheat HB4 commercialized by Trigall since January 2023 (see Chap. 5) provides a response to global warming with yields 15–20% higher in water stress situations.

Brazil started GM crops in 2003, later than Argentina, but has since supplanted it with 51.3 million ha devoted to biotech crops (more than half of the arable land) in 2018 and 52.8 million in 2019, an increase of 3% representing 28% of the world's biotech area. Brazil ranks just behind the United States (Table 12.1). Soybeans occupy 2/3 of the biotech acreage (35.1 million ha in 2019), with the remaining 1/3 shared between corn (16.3 million ha) and cotton (1.4 million ha). The percentages of transgenic plants are very high in these three crops: 97% for soybeans, 91% for corn, and 90.5% for cotton. They are mostly herbicide-tolerant (HT), but gene stacking is increasing (75% of corn grown in 2017 is IR/HT and 59% for cotton). GM sugarcane appeared in 2018 on 400 ha, and a year later, it is grown on 18,000 ha. Among the 68 transformation events that have been approved in this country, there are now several stacks that give soybeans double tolerance to two different herbicides (glufosinate and dicamba) or, accompanying two insect resistance genes, triple tolerance to three different herbicides (glyphosate, glufosinate, and 2,4-D), allowing for variation in herbicide treatments and therefore delaying the onset of resistance in weeds (see Chap. 11). It should be noted that 80% of Brazilian soybean exports go to China.

Paraguay and Uruguay are the other two Mercosur countries. They are small in size but major in terms of biotech agricultural crops: they are both among the countries growing biotech crops on more than 1 million ha:

- In Paraguay, glyphosate-tolerant HT soybeans (*Roundup Ready®*), authorized since 2004, are the dominant GM crop, occupying, in 2019, 87% of the cultivated biotech surface area (3.6 million ha), far ahead of corn (12% or 500,000 ha) and cotton (marginal 18,000 ha). These crops, which involve 10,000 farms, occupy 62% of the country's arable land. Paraguayan soybeans are 99% transgenic with a development, in recent years, of HT/IR stacks that goes from 17% in 2016, 35% in 2017, and 43% in 2019. This strong increase also underscores that farmers are satisfied with these versatile seeds. IR/HT multipurpose corn is popular with a 50% increase in acreage between 2018 and 2019 (at the expense of HT corn, no IR corn is planted) and represents 91% of the acreage of this crop.

- In Uruguay, 3000 farms are developing biotech soy and corn. Soybeans account for almost all the biotech areas cultivated (1.21 million ha). However, it is noting the strong increase in corn acreage: from 50,000 ha of corn grown in 2018 to 130,000 in 2019 with an adoption rate of 97% (10,000 ha HT and 107,000 ha stacked IR/HT) (85.7% adoption rate in 2016).

Other Latin American countries (Bolivia, Mexico, Colombia, Honduras, Chile, and Costa Rica), beyond biotech soybeans, corn, and cotton plants, are developing on small areas flax, rice, pineapple (*Pinkglow*™ with high added value in Costa Rica), and also blue carnations in Colombia for export to Japan. GM crops are progressing significantly in several countries: in 2019, they occupy 41,083 ha in Chile, 297 ha in Costa Rica, 223,000 ha in Mexico, 37,386 ha in Honduras, and 101,188 ha in Colombia [1], which represents considerable increases in surface areas in some countries between 2018 and 2019 ± 114% in Costa Rica, +44% in Chile, +15% in Colombia, and +8% in Bolivia.

The American continent is by far the one that has the most adhered to the culture of biotech plants.

Asia-Pacific Region: Significant Growth

This part of the world cultivates 10.2% of the world's biotech crop area in 2019, a 2% increase. India is the leading country with 11.9 million ha of cotton, followed by China with 3.2 million ha of cotton and papaya; Pakistan with 2.5 million ha of cotton; the Philippines with transgenic maize; Australia with 614,000 ha of cotton, canola, and safflower; Myanmar (Burma) with 300,000 ha of cotton; Vietnam with 90,000 ha of cotton; Bangladesh with 1931 ha of Bt eggplant (cf. Chap. 11); and Indonesia with 2000 ha of drought-tolerant sugarcane. Thus, nine countries in Asia are growing biotech crops, including six countries in Southeast Asia and the Pacific zone that operate on small areas of less than 1 million ha.

China was the first country to commercialize transgenic plants in the early 1990s. Since 1997, Bt cotton (IR) has been grown to control lepidopteran pests, and GM papaya has been grown on a small area for human consumption. The papaya is resistant to *Papaya ringspot virus* (PRSV). These two crops in 2017 occupied 7130 ha for papaya with an adoption rate of the transgenic crop of 86% and 2.8 million ha for IR cotton with an adoption rate of transgenic cotton of 95%, almost all the cotton crop area. In 2019, the biotech crop area has risen to 3.2 million ha. There are 7 million farmers involved in these two crops. China has approved 64 transgenic events for import or cultivation of cotton, Argentinian canola, corn, papaya, petunia, poplar, rice, soybean, sugar beet, bell pepper, and tomato. The proactive policy of the Chinese authorities to promote biotechnology has resulted in the granting of 3 billion dollars (USD) in funding to Chinese research institutes and companies to develop biotechnological lines of drought-resistant wheat and corn, disease-resistant rice, and soybeans with a nutritionally improved oil profile and increased yields.

India developed genetically modified cotton plantations to resist insects as early as 2002, but it was from 2007 onward that these crops became the majority with an adoption rate of 60% growing to 93% in 2017. With an area of 11.9 million ha in 2019, India is now the Asian country with the largest biotech crop area. However, the situation of transgenic plants throughout the country is complex and confusing. Indeed, this country is composed of states that have not all adopted the same position toward GMOs and face intense lobbying from powerful anti-GMO movements (see Chap. 9). Experimental trials to diversify crops have, however, been carried out with several species: chickpea, mustard, rice, sugarcane, and eggplant. Within the framework of partnerships with several Southeast Asian countries (Bangladesh, Indonesia, the Philippines), several research programs and field trials of delayed ripening papaya, insect-resistant cotton, and nutritionally improved rice are being developed.

Neighboring Pakistan is growing 2.5 million ha of various IR cotton varieties (34 were approved between 2010 and 2016) and is also developing trials of HT or stacked IR/HT corn.

In Myanmar which has recently adopted the cultivation of transgenic cotton plants, the adoption of biotech cotton plants is reportedly gaining momentum following the example of the cultivation of brinjal Bt eggplant in Bangladesh. *Ngwe chi-6* and *Ngwe chi-9* cotton varieties were planted in 2019 on 300,000 ha, occupying 85% of the total area planted with cotton in that country (350,000 ha). Burmese farmers are reportedly enjoying the yields obtained (2000 kg/ha) from these local varieties developed by the *Lungyaw Cotton Research Farm* and the MOALI (*Ministry of Agriculture, Livestock and Irrigation) Technology Development Farm.*

African Continent: After Doubts, Hope

Many trials are underway in West Africa (Burkina Faso, Cameroon, Ghana, and Nigeria), and also in East Africa (Eswatini – formerly Swaziland – Kenya, Mozambique, Malawi, Uganda) with crops of cowpea, maize, sorghum, cotton, sweet potato, rice, banana, and soybeans, but only three countries were growing biotech plants in 2018: Sudan with 243,000 ha of IR cotton, historically South Africa with 2.7 million ha, and a third country, Eswatini, with 250 ha of IR cotton planted in 2018.

In 2019, the African countries growing biotech plants doubled from three to six. Indeed, this trio has been joined by three other countries for the cultivation of Bt cotton, Malawi (6000 ha), Nigeria (700 ha), and Ethiopia (311 ha), and soon Kenya where a registration has been granted by the authorities. Eswatini has increased its Bt cotton acreage (401 ha), Sudan has increased its Bt cotton acreage (236,000 ha), and South Africa is growing 1.95 million ha of HT and IR maize, 693,975 ha of HT soybeans, and 43,654 ha of HT and IR cotton (Bt). It should be noted that Nigeria has just authorized the cultivation of transgenic cowpeas.

In all, 11 African countries are conducting research or have authorized ten biotech crops including banana, plantain, corn, potato, rice, sorghum, soybean, cowpea, cotton, and tapioca. Sixteen transformation events are under study or already commercialized: resistance to various crop pests and pathogens, drought tolerance, and biofortification (tapioca and sorghum).

This dynamic follows a long period of doubt linked to several circumstances. Indeed, several transgenic crops were grown in African countries and then abandoned: in Burkina Faso (IR cotton) or in Kenya (a project of herbicide-tolerant cotton).

- In Burkina Faso, the reason given was that some cotton varieties that had incorporated the Bt transgenic event to control the pink cotton worm (*Pectinophora gossypiella*) had the defect of producing cotton fibers that were shorter, of poorer mechanical quality, and less adapted to handpicking than conventional African varieties. Despite the reduction in pesticide treatments (to one third) and the increase in yields (double), the commercialization of the harvest of this cotton was less lucrative for the cotton companies, which have the upper hand on the seeds that they supply to Burkinabe farmers, and they abandoned these varieties that did not correspond to their expectations.
- The *Séralini affair*[2], with the publication of photos of rats deformed by tumors on the front page of newspapers, has left its mark in Kenya, even if the incriminated research article [3] has since been invalidated by the scientific community and retracted by the journal that published it.

Claudine Franche, honorary Director of Research at the *Institut de Recherche pour le Development* (IRD), describes the African hesitations in these terms:

> This is due to a number of factors, including the lack of political will to put in place the necessary legislative and legal framework, the risk of seeing the agricultural sector under the control of large multinationals, the lack of scientific expertise allowing for an in-depth benefit-risk analysis of these crops, and finally the influence, on this continent, of Europe where the precautionary principle prevails [4].

Among the projects under development in Africa, the WEMA (*Water Efficient Maize for Africa*) project, which aims to adapt maize cultivation to drought, is an exemplary international project in terms of the international cooperation that has been established around it. It is the result of a public-private partnership between CIMMYT (*Centro Internacional de Mejoramiento de Maiz y Trigo/International Maize and Wheat Improvement Center*), which provides high-yielding maize varieties adapted to African conditions, and national agricultural research organizations

---

[2] The study by M. Kuntz published by the Fondation pour l'innovation politique [2] is worth consulting.

in Kenya, Mozambique, Uganda, South Africa, and Tanzania, which provide development infrastructure, as well as Monsanto (now Bayer), which contributes its knowledge of the MON 87460 transgenic event in maize varieties. First grown in the United States in 2013 and in South Africa in 2014, varieties of this transgenic corn have yields that are 7–15% higher under water stress conditions. Tests to develop gene stacks (insect-resistant, better nitrogen use) are continuing in the African partner countries of this project. Funded by the *Bill & Melinda Gates Foundation* and the *Howard G. Buffett Foundation* (HGBF) and with the *United States Agency for International Development* (USAID), the WEMA project aims to make royalty-free seeds available to African farmers. The TELA project (from the Latin *tutela* meaning protection) is following the WEMA. Its goal is to develop drought-tolerant and insect-resistant maize varieties through public-private partnerships [5] for royalty-free seeds.

The Iberian Exception in a Hostile European Continent

From Russia, which renewed its ban on the cultivation and import of GMOs in 2020 as part of the new food security doctrine, to the European Union, which has allowed a policy of distrust of GMO cultivation to take hold, supported by political parties with political calculations or collapsological convictions[3] (see Chap. 5), the European continent is at best reluctant, at worst hostile, to the cultivation of transgenic plants.

Only two countries in the European Union (EU) continue to cultivate GMO plants in a not very encouraging community environment: Spain and Portugal.

In 2019, the Iberian Peninsula hosts fields of GMO maize Bt MON 810, the only transgenic plant grown today in the EU, on an area of 111,883 ha (120,990 ha in 2018, 131,535 ha in 2017), that is, 107,130 ha in Spain and 4753 ha in Portugal. This decrease in the area cultivated with maize is explained not only by the very strong psychological pressure in the European Union against GM crops but also and above all by the imports of maize into Spain from South America and the United States, which exert very strong competition.

MON 810 maize is a Bt maize (insect-resistant (IR)) that has been grown for more than 15 years in Spain without incident and to the great benefit of the farmers who plant it and whose agronomic and sanitary advantages have been recalled above (see Chap. 11). It saves insecticide treatments, and its crops contain less mycotoxins. In fact, it is in the regions where the European corn borer (*Ostrinia nubilalis*) and the pink stem borer (*Sesamia nonagrioïdes*), major insect pests of the crop, are endemic and exert a strong parasitic pressure, in Aragon and Catalonia and to a lesser extent in Extremadura, that the areas cultivated with MON 810 are concentrated (82% of the total area). From 1 year to the next, the farmer chooses to

---

[3]Collapsology is a current of thought that considers as inevitable a global and systemic collapse of the western industrial civilization as a result of a combination of environmental, energetic, economic, geopolitical, democratic crises, etc.

grow transgenic corn or to return to conventional varieties if the pest pressure has temporarily decreased, as was the case in Catalonia in the 2010s.

In Portugal, MON 810 is grown in dedicated regions (Alentejo and the Lisbon region) where the structure of farms is not an obstacle to the European regulation on the coexistence of agricultures (see Chap. 11).

Today, a large majority of EU countries reject the cultivation of biotech plants, after having accepted it, for various reasons linked to the different national contexts and especially to the relentless action of anti-GMO NGOs. However, it should be noted that although the EU only cultivates very marginally biotech plants, it imports them massively: it is, after China, the second largest importer of soya in the world, 80% of which is transgenic!

A Technology Adopted by Developing Countries and Poor Farmers

This technology was rapidly adopted not only by developed countries but also by developing countries, which matched and then surpassed them in cultivated area. In 2010, the progress curves of these countries converged; 8 years later, developing countries are growing more than 100 million ha of biotech crops (54% of the world's biotech area), and every year, the biotech land cultivated in these countries is increasing significantly. The progress of biotech agriculture is therefore greater in emerging and developing countries.

This finding is very consistent with the fact that, contrary to popular belief, it is not the large farms of North America or the *haciendas* and *fazendas* of Latin America that have the privilege of growing transgenic crops, but millions of small farmers, particularly in China and India. Their number has increased from 12 million to 17 million between 2007 and 2018, with 90% of farmers having very low incomes. Among the arguments put forward by these small farmers are that their work is less arduous, that their health is better protected by reducing the painful and frequent spraying of phytosanitary products on their crops, and that their incomes are better because of the increase in yields, as highlighted by the textbook case of Bt eggplant cultivation in Bangladesh (cf. Chap. 11).

Commercialization Now Under the Aegis of International Consortia

At the beginning of the cultivation of biotech plants at the end of the 1990s, the commercialization of transgenic plants was carried out by a majority of the large private companies of the time (Monsanto, AgrEvo, Florigene, Calgene, Zeneca, etc.) but also by universities that valorized the research work of their laboratories (e.g., Cornell University, University of Saskatchewan) [6]. At that time, the large conglomerates as we know them today did not exist. But the companies involved in the development of biotechnologies quickly understood the strategic interest of getting closer to the seed and agrochemical companies.

The American company Monsanto has become, in spite of itself, the symbol of these groupings (Box 12.1) and has acquired, because of its dynamism and industrial success, the unenviable status of being the scapegoat of the anti-GMO movement.

Over the years, numerous mergers and acquisitions of players in these sectors have taken place, shaping the current structure (Box 12.2 and Fig. 12.1). These consortia have the means to deal with increasingly demanding regulations, particularly those of the European Union, which require very costly applications for approval and post-marketing follow-up.

Box 12.1: Monsanto: An Extraordinary Industrial Epic (1901–2017)

Monsanto is an American company founded in 1901 by John Francis Queeny, whom he named after his wife: Olga Mendez Monsanto [7]. Before the First World War, it produced saccharin, caffeine, and vanillin for its client, a young company, and then in full development Coca-Cola! After the war, it diversified by producing aspirin from 1918, and Monsanto remained the leading American producer until the 1980s.

From 1929, the company moved into the rubber and phosphate sectors. In 1950, in partnership with the American companies *American Viscose* and *Nylon*, it created Acrilan acrylic fibers, a product of acrylonitrile polymerization.

At the same time, starting in 1945, it began producing herbicides and insecticides, including 2,4-dichlorophenoxyacetic acid (2,4-D) and 2,4,5-trichlorophenoxyacetic acid (2,4,5-T), two defoliating agents that were used in mixtures by the American army during the Vietnam War (1961–1971). This pink-brown mixture is better known as *Agent Orange*. The damage caused to the Vietnamese population tarnished the reputation of the two companies that supplied the American army, Monsanto and Dow Chemical.

In 1970, a company chemist, John E. Franz, synthesized glyphosate, which was marketed in 1974 under the name *Roundup®*. At the same time, in 1978, Monsanto invested in molecular biology programs, and one of its teams, led by Robert Fraley, created a transgenic petunia in 1983 (see Chap. 11), followed by the commercialization of many other genetically modified plants: the *NewLeaf* potato, *Bollgard* cotton plant, *YieldGard* corn, *Roundup Ready* soybeans, *Roundup Ready* oilseed rape, and *Roundup Ready* cotton, the last three being glyphosate-tolerant plant varieties. Thus, Monsanto owns the patents on the transgenic seeds and on the herbicide that is to be used on these seeds: high commercial strategy!

(continued)

Box 12.1 (continued)

The company was then at the heart of a whirlwind of mergers and acquisitions. The company, which had always maintained a range of activities centered on the nutrition, health, and agriculture triptych, after having sold its plastic materials division in the 1990s, merged with pharmaceutical giant Pharmacia & Upjohn in December 1999 and became the Pharmacia company. However, it was decided to separate the pharmaceutical and agricultural divisions. In 2002, the agricultural business became independent and kept the name Monsanto.

In 2005, Monsanto bought Seminis Inc. after many other companies in the sector, thus becoming the world's largest seed company. Although after 20 years its patent on glyphosate has fallen into the public domain, the industrial group symbolizes the omnipotence of the alliance that has been created between the seed and agrochemical sectors. It becomes the target of anti-GMO and environmentalists.

In 2017, it was Monsanto's turn to be bought by the German group Bayer, which decided to abandon the American brand loaded with too strong a symbolism.

Box 12.2: Large International Conglomerate Leaders in the Agrochemical-Biotechnology Sector

Mergers and acquisitions between companies in the agrochemical and biotech sectors over the past 25 years have resulted in the creation of three large international conglomerates headquartered on three different continents: in the United States for Corteva, which resulted from the merger between DuPont de Nemours and Dow AgroSciences; in Asia for ChemChina, which absorbed the Swiss-based company Syngenta; and in Europe for Germany's Bayer, which acquired the US-based Monsanto. The latest restructurings occurred between 2017 and 2019 (Fig. 12.1). Together with the BASF Group, these consortia dominate the agrochemical and seed sectors. Today, all of these agribusiness groups have a range of biotech and agri-supply companies (fertilizers, synthetic or biocontrol crop protection products) within them.

What Can We Learn from 25 Years of GMO Crops?

At the end of this overview, what are the landmarks that emerge from a quarter century of GMO cultivation? The global map of countries growing GM crops in 2019 (Fig. 12.1) highlights:

- The world is divided in two: On the one hand, the countries that grow and export biotech plants are located in North and South America, Asia and the Pacific zone (Australia, New Zealand), some African countries, and two EU countries (Spain

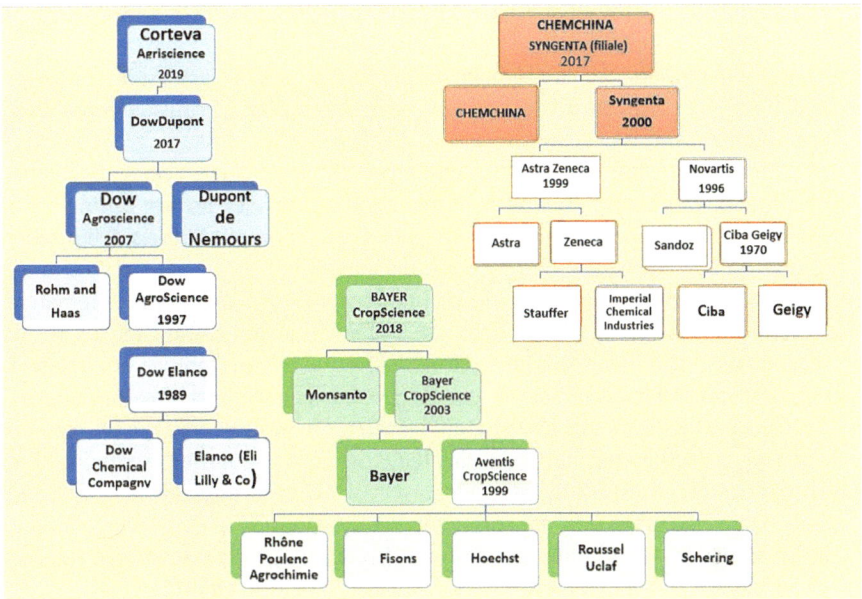

Fig. 12.1 Mergers and acquisitions to form the three major groups that, together with BASF, dominate the agrochemicals and biotechnology sector. (Source: Ambolet and Regnault-Roger [8, 9] adapted and updated)

and Portugal). On the other hand, the countries that do not cultivate but import them are Russia, the majority of African and EU countries, and the countries of the Middle East. This separation is old.

- Twenty-nine countries grow biotech plants in 2019 and 43 countries import them, so 72 countries use them.
- Biotech crops occupied 190.4 million hectares in 2019, a slight decrease of −0.7% compared to 2018 (191.7 million ha), which can be explained by climatic conditions (drought) in certain parts of the world. However, this is a global balance sheet that can be qualified by other elements (see below).
- Developing countries, which surpassed developed countries in area in 2010, continue to grow (+2.5% compared to 2018) with 105.7 million ha, while developed countries show a decrease in cultivated area with 84.7 million ha (or −4.4%), compared to 2018.
- The countries with increased GM acreages are Brazil, Paraguay, India, the Philippines, China, Bolivia with areas exceeding 100,000 ha (with a notable increase of 1.5 million ha between 2018 and 2019 in Brazil) and for smaller areas, Vietnam (+86% in 1 year), the Philippines (+39%), Colombia, Mexico, Honduras, Chile, Indonesia, and Bangladesh.
- The countries with a decrease in the area cultivated with biotech crops in the last year amounts to more than 100,000 ha are the United States, Pakistan, Canada, Australia, and Uruguay.

- Five of the 29 countries that grow GMOs do so on more than 10 million hectares: the United States, Brazil, Argentina, Canada, and India.
- 17 million farmers, 16 of them in developing countries, grow biotech crops, which together with their families represents more than 65 million people who earn income from biotech agriculture.
- The American continent (North and South) is still the land of choice for biotech plants, accounting for nearly 90% of the world's cultivated area in 2019.
- The African continent is now opening to GM crops after marked hesitation (*Séralini case*, Burkinabe Bt cotton). Alongside South Africa, a country that has historically grown them, six African countries are now developing commercial biotech crops (Nigeria, Sudan, Ethiopia, Malawi, Eswatini, and Kenya), mainly IR cotton, while Ghana, Burkina Faso, Uganda, and Mozambique are authorizing experimental trials of various crops (rice, cowpea, banana, potato, cassava).
- Southeast Asia shows that government resistance to biotech crops as the result of activism by influential anti-biotech NGOs in some countries is diminishing. Three examples illustrate this: the cultivation of Bt brinjal eggplant in Bangladesh and golden rice in the Philippines and the recent adoption of GM cotton varieties in Myanmar. Nine countries in the Asia-Pacific region have increased the area under biotech crops by an overall 2% or nearly 20 million ha (Fig. 12.2).

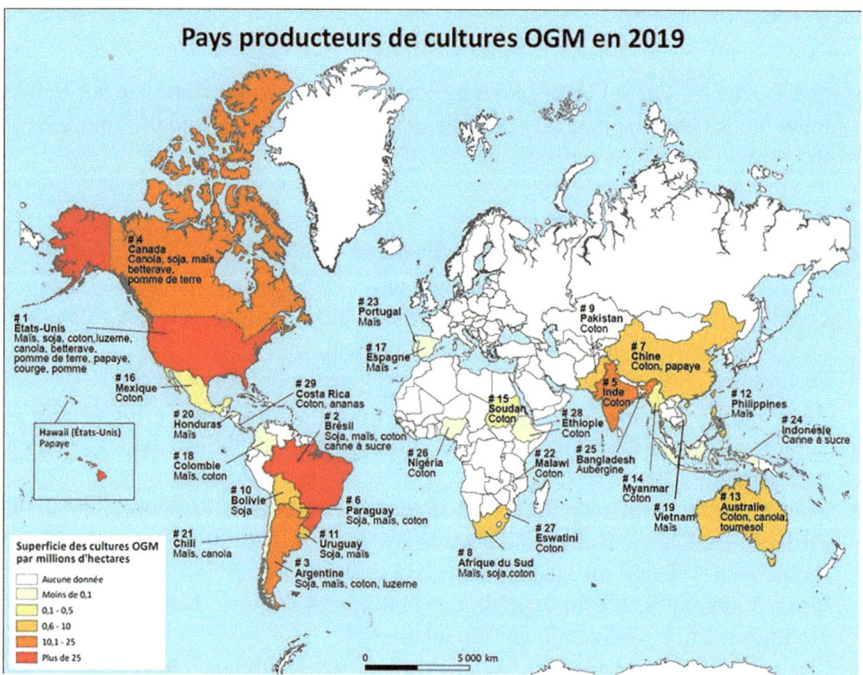

Fig. 12.2 World map of countries growing GMOs in 2019. (Source OGM.gouv.qc.ca (with courtesy))

Conclusion

Any development of new seeds, whether conventional or transgenic, requires significant research work that takes several years before they are marketed. The aim is to adapt the varieties to the climatic and environmental conditions of the different countries. In developing countries, this long-term work mobilizes cooperation between biotechnology companies, universities, or local specialized research institutes and public or private funding (such as the *Bill & Melinda Gates Foundation*, which is very involved in this field, but also the *Howard Buffett Foundation* or USAID (*United States Agency for International Development*)).

The rise of first-generation biotech crops (GMOs) in developing countries in Africa and Asia is part of this logic, while at the same time research in several developed countries is increasingly turning to second-generation biotechnologies and genome-editing techniques which are more recent and easier to develop.

Bibliography

1. ISAAA. (2018). Global Status of Commercialized Biotech/GM Crops in 2017: Biotech Crop Adoption Surges as Economic Benefits Accumulate in 22 Years *Brief #53*, https://www.isaaa.org/resources/publications/briefs/
2. Kuntz, M. (2019). *The Séralini affair, dead end of an activist science* (p. 60). Foundation for Political Innovation.
3. Séralini, G. E., Clair, E., Mesnage, R., Gress, S., Defarge, N., Malatesta, M., Hennequin, D., & Spiroux de Vendômois, J. (2012). RETRACTED: Long term toxicity of a roundup herbicide and a roundup-tolerant genetically modified maize. *Food and Chemical Toxicology, 50*(11), 4221–4231. https://doi.org/10.1016/j.fct.2012.08.005
4. Franche, C. (2018). Biotechnologies végétales et pays en développement. In C. Regnault-Roger, L. –M. Houdebine, & A. Ricroch (dir), *Au-delà des OGM* (pp. 115–134). Presses des Mines.
5. AATF. (2021). *TELA maize Project*, https://www-aatf%2D%2Dafrica-org.translate.goog/tela-maize-project/?_x_tr_sl=en&_x_tr_tl=en&_x_tr_hl=en&_x_tr_pto=sc
6. James, C., & Krattiger, A. F. (1996). Global review of the field testing and commercialization of transgenic plants, 1986 to 1995: The first decade of crop biotechnology. *ISAAA Briefs No. 1*, ISAAA: Ithaca, pp. 31, https://www.isaaa.org/resources/publications/briefs/01/download/isaaa-brief-01-1996.pdf
7. This historical summary is based on the article "Monsanto-Definition and explanation" Trch-sciences.net., https://www.techno-science.net/glossaire-definition/Monsanto
8. Regnault-Roger, C. (2020). *Des plantes biotech au service de la santé du végétal et de l'environnement* (p. 56). Fondation pour l'innovation politique.
9. Ambolet, B. (2018). La chimie organique au service de l'agriculture. In C. Regnault-Roger & A. Fougeroux (Eds.), *Santé du végétal: 100 ans déjà!* (pp. 77–88). Editions Presses des Mines.

Part III
New genomic Techniques (NGTs) What Prospects? What Are the Challenges?

Chapter 13
NGT: In the R&D Stage

New genomic techniques (NGTs) are revolutionizing therapeutic approaches in human medicine, as seen in the management of the current COVID-19 pandemic, with new vaccines providing effective new tools and, in record time, to combat the SARS-CoV-2 coronavirus and its many variants. They also open new perspectives for improving veterinary medicine and animal welfare. Applications in the plant sector are not to be outdone.

In this section, we propose to provide an overview of current achievements and future projects developed on different continents in order to better understand the emerging biotechnological issues.

Numerous R&D Projects

The dynamism of a new technique is measured by the research work devoted to it and, in the current system of this sector, the publications in international-level journals.

When we search databases using the term CRISPR, few research projects on the subject were mentioned before 2012: with 145 publications listed by PubMed[1], it appeared that this was a niche research field like so many others. But since the article published in *Science* in August of that year by Emmanuelle Charpentier and Jennifer Doudna and their collaborators unveiling the full potential of this approach for targeted genome modification, the research work has taken off considerably. Researchers have adopted this technique in record time and have been very enthusiastic about it: in fact, up to April 2021 (over a period of almost 9 years),

---

[1] The leading search engine for bibliographic data in all areas of biology and medicine.

© The Author(s), under exclusive license to Springer Nature Switzerland AG 2023
C. Regnault-Roger, *Biotech Challenges*,
https://doi.org/10.1007/978-3-031-38237-6_13

Table 13.1 Number of R&D projects under development with NGTs and application areas (based on elements given in the JRC report [2])

Fields of application	Beginning R&D	Advanced R&D
Medical/therapeutic	58	61
Animal	63	32
Plants and mushrooms	292	117
Total	413	210

no less than 23,838 scientific publications have been counted. More than three quarters of the articles related to NGTs are dedicated to it [1].

This enthusiasm can also be seen in the range of research and development (R&D) projects underway, which illustrates the strong development potential of these techniques. For example, the European JRC study published in spring 2021 [2] counted 623 research and development (R&D) projects involving NGTs at various stages of development for commercialization in the plant (66.5%), animal (15%), and medical (18.5%) fields (Table 13.1). Most of them are in the early stages of research (64%) and 36% in a more advanced phase, 18 are even in a very advanced pre-commercialization phase. They are currently being marketed or are expected to be marketed by 2030.

The CRISPR technique was chosen to develop 68% of the projects, which supports the analysis that most of these R&D projects are very recent since the technique was invented just 10 years ago.

Patents Mainly in the United States and China

The vitality of a scientific research is also manifested by the number of patents that are granted, reflecting the innovative efforts. Examining the geographical origin and applications of patents filed on the CRISPR technique between 2013 and 2018 is very revealing of the orientations and choices made in the different countries (Fig. 13.1).

It highlights the dynamism of Chinese and American laboratories, which together own nearly 80% of the patents. The two most inventive countries are undoubtedly China and the United States. But if American laboratories remain leaders in technical improvements and in the medical application sector, it is China that has filed the most patents in the industrial and agricultural (plant and animal) application sectors [4].

One can only note sadly that the European Union has become quite marginal in this sector and is at the same level as all the other countries in the world combined, far behind the two leaders. Who remembers that, in the 1990s, the European continent dominated innovation in biotechnology and "led the way" (see Chap. 2) because European companies had filed more patents on biotechnology than American companies? Today, 30 years later, it is the United States and China that are waltzing together and dominating the world! For the year 2020 alone, the

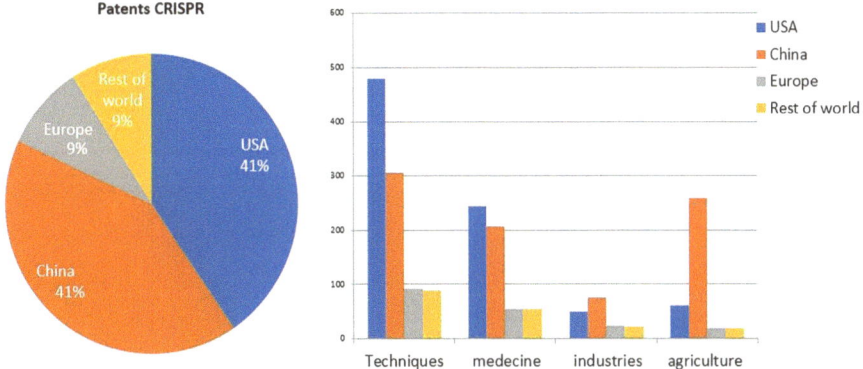

Fig. 13.1 Geographical distribution of patents taken on CRISPR and fields of application. (Source: Regnault-Roger [3])

number of patents filed in plant biotechnologies is 10,624 for China and 8800 for the United States, while France only produces 1007, Germany 2048, Japan 2143, Russia 588, and India 117 according to the *World Intellectual Property Organization* (WIPO) [5].

This European delay in scientific production is accompanied by a real rout in terms of intellectual property. Regulation, not only for GMOs but also for NGTs, is now a real geopolitical issue and a matter of economic independence.

Bibliography

1. Chneiweiss, H., & Hirsch, F. (2021, April 9). From health to food, the applications of genome editing. *The Conversation*, https://theconversation.com/de-la-sante-a-food-the-applications-of-genome-editing-158538
2. Parisi, C., & Rodriguez-Cerzo, E. (2021). *Current and future market applications of new genomic techniques* from Joint Research Centre EUR 30589 EN, p. 49.
3. Regnault-Roger, C. (2020). *GMOs and genome-editing products: Regulatory and geopolitical issues* (p. 56). Foundation for Political Innovation.
4. Martin-Laffon, J., Kuntz, M., & Ricroch, A. (2019). Worldwide CRISPR patent landscape shows strong geographical biases. *Nature Biotechnology, 37*, 601–621.
5. WIPO/WIPO Databases cited by Regnault H (2022). La ménagerie impériale. *La Crise* n. 42 (January 2022), p. 8, CEIM, UQÀM, https://www.wipo.int/portal/en/

Chapter 14
Second-Generation Medical Biotechnologies: High Expectations

The COVID-19 pandemic has sparked a race in research laboratories around the world to understand the biology of this virus and find solutions to limit its effects. Apart from a minority anti-vax movement, an attitude that has recurred throughout the ages as underlined by the winners of the Villemot Prize of the Academy of Sciences 2020 [1], who today would refuse to use first- and second-generation biotechnologies to find a response to the SARS-CoV-2 coronavirus, the agent of COVID-19? Why, when there are great hopes for the treatment of rare or orphan diseases, for example? The NGTs open many prospects.

Therapeutic Hopes

In human health, NGTs hold out hope for treating several hereditary genetic diseases or conditions with poor prognosis or for alleviating the lack of organ donations to treat patients in need of transplantation.

Here are two examples:

- In 2017, a parliamentary mission of the OPECST had mentioned the gene therapy using the TALEN genome-editing technique implemented on two 11- and 18-month-old girls with acute lymphoblastic leukemia (a cancer of the white blood cells with a poor prognosis) and treated at *Great Ormond Street Hospital* in London in 2015 [2]. The girls are now doing well.
- Experimental transplantation of a CRISPR-edited pig kidney into a human was successfully performed at *Langone University Hospital* (New York University) on October 20, 2021 (Box 14.1).
  Genome editing thus paves the way in human medicine for the transplantation of animal organs. The number of patients requiring organ transplants is growing faster than the availability of human organs. Pig organ transplantation could be a

palliative if the many antigens that trigger rejection with varying intensity in the patient are neutralized. The use of the CRISPR/Cas9 technique can significantly reduce the porcine antigens responsible for rejection of pig organs in humans.

> **Box 14.1: Xenograft and NGT**
> Professor Robert Montgomery's successful experiment at Langone Hospital was conducted with a pig called *GalSafe* genetically modified at the embryonic stage to have a gene that eliminates the presence of a sugar, alpha gal (galactose-alpha-1,3-galactose). This sugar, which does not exist in humans, causes allergies and rejection of animal transplants that contain it. The aim is to develop xenografts [3]. The *GalSafe* pig marketed by *United Therapeutics* was approved by the US health regulator, the *Food and Drug Administration* (FDA), in late 2020 [4]. Another goal of *United Therapeutics* is to produce therapeutic molecules such as heparin, an anticoagulant [5].

Ongoing Developments in Medical Research

Providing an Answer to Orphan Diseases

The therapeutic applications of NGTs cover many pathologies:

- Hereditary genetic diseases (hereditary deafness, cystic fibrosis, congenital Kabuki syndrome, which results in psychomotor retardation, Rubinstein-Taybi syndrome caused by genetic mutations and resulting in mental retardation, Huntington's disease, which affects adults between the ages of 30 and 50 and results in motor, behavioral, and psychiatric disorders) and neurodegenerative diseases that are common among the elderly (Alzheimer's, Parkinson's). They have given rise to 20 projects, 4 of which are at the clinical trial stage. Hemoglobinopathies (sickle cell disease, thalassemia, hemophilia) and cardiovascular and metabolic diseases (obesity, diabetes) are the subject of 20 studies, 10 of which are in clinical trials. In total, in the spring of 2021, 64 clinical trials are in progress, in either phase I or phase II. Most of these trials began between 2018 and 2020.
- Cancers of various etiologies[1] (gastrointestinal, lung, prostate, pancreatic, ovarian, renal carcinoma, neurofibromatosis, myeloma and melanoma, leukemia, lymphoma) have given rise to 48 studies, some of which are at an advanced stage, since 30 clinical trials have been carried out (5 of which have already been withdrawn for noncompliant results). Some cancers are induced by viruses (papillomaviruses that infect the genital tract and cause cancers of the vagina, cervix, or penis and the Epstein-Barr virus (EBV) that causes several diseases such as

---

[1] Etiology: study of the causes of diseases.

infectious mononucleosis). These are the subject of eight projects, five of which are already in clinical phase.

- Viral diseases (HIV (*human immunodeficiency virus*) of AIDS, coronaviruses of COVID-19, noroviruses that cause gastroenteritis) to which 23 studies are dedicated with 11 in clinical phase and 12 in laboratory research stage.
- Some very successful projects have already received FDA approval: for example, a gene therapy to lower LDL (*low-density lipoprotein*) cholesterol levels in the blood was approved in 2013. The treatment of a rare hereditary disease, hereditary retinal dystrophy due to a mutation in the *RPE65* gene (major vision problems), has been developed by gene therapy. It consists of injecting a viral vector containing a functional copy of the altered *RPE65* gene: this healthy gene is delivered to the cells of the retinal pigment epithelium and allows to slow down their degeneration. This gene therapy was approved in 2017 by the FDA and by health authorities in several countries: the University Hospital (Retina Institute) in Nantes is authorized to provide this treatment [6].
- Studies are also being conducted on stem cells. For example, cells from patients with Wolfram syndrome (WFS1), an insulin-dependent diabetes that results from a defective gene and leads to a short life expectancy, have been modified by CRISPR to correct the genetic error and produce insulin. The *University of Washington School of Medicine* team leading this work also transformed pancreatic β cells with CRISPR to secrete more insulin, then implanted them under the skin of mice made diabetic, and cured them. Clinical trials are winding down. These experiments open the door to gene therapies to treat diabetes [7, 8]. One study aims to improve the tolerance of kidney transplants by replacing the administration of conventional immunosuppressants, which over the long term become toxic for the patient and cause cardiac disorders, diabetes, or an increased risk of infection, with genetically edited stem and immune cells [9].

Among the 119 R&D projects in progress, the CRISPR technique is used in the majority (55%) of projects, while the ZFN technique (zinc finger nucleases) is used in 30% of cases, the TALEN technique in 13%, and a few studies conducted with RNAi. Some of this research has been going on for a long time even before the discovery of the CRISPR technique.

Development of Diagnostic Tools

NGTs are also used for the development of diagnostic tools. Five projects are under development [10]:

- Three are performed in China. Two are on pulmonary pathologies (severe sepsis[2] and tuberculosis). The third one concerns the diagnosis of COVID-19 and is

---

[2] Generalized inflammatory response associated with severe infection.

considered a major national research program. These researches are conducted in laboratories all over China, especially for pulmonary researches, in different hospitals: *Huashan Hospital* in Shanghai, *Wenzhou Central Hospital*, and the *Red Cross Hospital* in Hangzhou.

- Two projects, developed in the United States, are aimed at diagnosing ovarian cancer and neurofibromatosis type 1 (Recklinghausen's disease), a monogenic disease due to spontaneous mutations (50% de novo mutation) that has a prevalence of 1/5000. The severity of the disease is highly variable from one individual to another. This study is being conducted at *Children's National Hospital* in Washington, D.C., a pediatric acute care children's hospital affiliated with both *George Washington University* (*School of Medicine*) and *Howard University* (*College of Medicine*).

Vector-Borne Disease Control

NGTs also open up new perspectives with regard to vector-borne diseases. The CRISPR technique associated with *gene drive* makes it possible to fight mosquitoes that carry infectious diseases (dengue, malaria, chikungunya, Zika), which are no longer confined to tropical areas but are now spreading throughout the world due to climate change and intercontinental travel. In addition to these mosquitoes, applications have also been extended to invasive animal species (proliferation of rats and feral cats[3]). However, a global consensus on the use of this approach, which can lead to local eradication of species, has yet to be reached [11].

Biofortified Foods

NGTs can also be used to modify the nutritional composition of foods in order to transform them into nutraceuticals.

Two applications have just been commercialized in Japan and the United States. The US company *Calyxt* has edited a soybean to produce soybean oil with no transfatty acids and less saturated fatty acids, which improves its nutritional quality. This modified soybean oil has received approval for marketing in the US market in 2019. In Japan, a tomato biofortified with high concentrations of γ-aminobutyric acid (GABA), a neuromodulatory amino acid that inhibits the prolonged excitation of neurons (relaxing effect and reduction of blood pressure), has been approved for marketing. This tomato, named *Sicilian Rouge High GABA*, is

---

[3] Hare cats (stray cats) or feral cats are domestic cats that have returned to a wild state and live in colonies.

presented by the *University of Tsukuba* and its start-up *Sanatech Seed*, its developers, as a health food. Its seeds are now in Japanese garden centers, and the first *GABA* tomatoes were consumed in front of Japanese television cameras in early January 2022!

In Spain, a research team at the Institute of Sustainable Agriculture (IAS, (*Instituto de Agricultura Sostenible*)) of the Superior Council of Scientific Research (CSIC (*Consejo Superior de Investigaciones Científicas*)) in Cordoba implemented the CRISPR/Cas9 technique to inhibit 35 of the 45 genes involved in the synthesis of gliadins, proteins responsible for allergies to wheat gluten. The low gluten content of these wheats could provide relief to patients with celiac disease [12]. Clinical trials are currently being conducted with gluten-reduced bread.

These few examples demonstrate that NGTs are already being used to provide food-based solutions to prevent nutritional diseases or improve human well-being. Designing nutraceuticals has become easier.

Two Countries Dominate

Many projects are being carried out in the United States, which is not surprising, since this country holds the largest number of patents based on CRISPR technology for medical applications, followed closely by China, while other countries are far behind (see Chap. 13). In fact, it is the United States that is developing the most research in the medical and therapeutic sector. About 900 projects based on NGTs are currently in the R&D pipeline, at the stage of laboratory studies and clinical trials. Work on sickle cell disease reached the stage of treating one patient, a 34-year-old woman, in the United States in 2019 [13]. China is also conducting a great deal of research to prevent or cure various cancers.

This dynamism in the medical and pharmaceutical sectors is reflected in the number of patents held by each country. In 2020, the WIPO (*World Intellectual Property Organization)* attributed 26,411 patents for medical technologies to the United States and 10,561 to China and 12,205 for pharmaceutical products to the former and 8722 to the latter. These figures should be compared with the patents held by Japan (9015 for medical technologies and 2662 for pharmaceuticals) and Germany (4572 and 1966 respectively) [14].

This spirit of innovation must however respect an ethical code, especially in the field of human reproduction.

Work in China has been singled out for research involving the manipulation of human embryos (since born) with CRISPR/Cas9 to prevent them from infection with the human immunodeficiency virus (HIV) that causes AIDS. Dr. He Jiankui, who presented these controversial findings at the *Second International Genome Editing Summi*t in Hong Kong in November 2018, was strongly condemned, figuratively by the entire scientific community and literally by Chinese authorities to a 3-year prison sentence.

In the face of international concern, the WHO has taken up the subject to define ethical rules excluding genome modification work on human germ cells and to set up a central register for monitoring research on modifications to the human genome in order to provide a powerful framework. Dr. Soumya Swaminathan, WHO Chief Scientist, said that "The committee will develop essential tools and guidance for all those working on this new technology to ensure maximum benefit and minimal risk to human health" [15].

Globalized Medical Research and Partnerships

These numerous studies highlight the dynamism of the medical scientific community in both private and public sectors. Numerous synergies exist between universities and public research institutes and companies, as demonstrated by the partnerships created to develop anti-COVID vaccines between the company *Moderna*, and the NIH (*National Institutes of Health*) in the United States and in Europe between *Oxford University* and the Anglo-Swedish company *AstraZeneca*.

These public-private partnerships also exist in the European research framework programs. Two framework programs FP7 (2007–2014) and H2020 (2014–2020) have thus funded 802 basic and applied medical research projects involving new genomic techniques (NGTs) for a total of 2.5 billion euros [16]. The subjects of the studies presented indicate that the scientific community has widely approved the CRISPR technique, which has become the flagship technique of NGTs. Indeed, if the projects concerning it represented only 27% of the FP7 projects for a funding of 700 million euros (the technique was published in 2012 and the program ended in 2014), it is the basis of 87% of the H2020 projects for an amount of 1.8 billion euros, that is to say, a 2.5-fold increase in the credits devoted to it.

Conclusion

Biotechnology is one of the tools widely used to advance human health, and NGTs, especially CRISPR, have been rapidly adopted.

The range of applications and the prospects for curing single-gene diseases and certain cancers are driving human medicine forward in leaps and bounds. This is demonstrated by the advances that have been made in addressing the COVID-19 pandemic, both in terms of understanding the disease and in developing breakthrough vaccines.

Bibliography

1. Salvadori, F., & Vignaud, L.-H. (2019). *Antivax, la résistance aux vaccins du XVIIIᵉ siècle à nos jours* (p. 347). Vendémiaire.
2. Le Déaut, J. -Y., & Procaccia, C. (2018). Une réflexion parlementaire pour l'avenir. In C. Regnault-Roger, L. M. Houdebine, & A. Ricroch (dir), *Au- delà des OGM* (pp. 181–196). Presses des Mines.
3. Dolgin, E. (2021). First GM pigs for allergies. Could xenotransplants be next? *Nature Biotechnology, 39*, 397–400. https://doi.org/10.1038/s41587-021-00885-9
4. Cooney, E. (2020, December 14). FDA approves genetically altering pigs, to potentially make food, drugs, and transplants safer. *Statnews*, https://www.statnews.com/2020/12/14/fda-approves-genetically-altering-pigs/
5. Deluzarche, C. (2021, October 24). World's first xenograft of a pig's kidney onto a human. *Futura santé*, https://www.futura-sciences.com/health/news/organ-first-world-xenograft-pig-kidney-human-6842
6. Institut de la rétine CHU de Nantes. (accès 2021). *Dystrophies héréditaires de la rétine*, Retina Institute Nantes University Hospital, https://www.retine-chirurgie-nantes.fr/dystrophies-hereditaires-de-la-retine
7. Maxwell, K. G., Augsornworawat, P., Velazco-Cruz, L., Kim, M. H., Asada, R., Hogrebe, N. J., Morikawa, S., Urano, F., & Millman, J. R. (2020). Gene-edited human stem cell-derived β cells from a patient with monogenic diabetes reverses preexisting diabetes in mice. *Science Translational Medicine, 12*(540), eaax9106. https://doi.org/10.1126/scitranslmed.aax9106
8. Washington University School of Medicine. (2020). *Diabetes reversed in mice with genetically edited stem cells derived from patients*, https://medicalxpress.com/news/2020-04-diabetes-reversed-mice-genetically-stem.html
9. Doheny, K. (2019, August 12). Stem cells and health advances: Where are we now. *WebMed*, https://www.webmd.com/brain/news/20190812/stem-cells-and-health-advances-where-are-we-now
10. Parisi, C., & Rodriguez-Cerzo, E. (2021). *Current and future market applications of new genomic techniques*, Joint Research Centre EUR 30589 EN, p. 49.
11. Kelsey, A., Stillinger, D., Binh Pham, T., Murphy, J., Firth, S., & Carballar-Lejarazu, R. (2020). Global governing bodies: A pathway for gene drive governance for vector mosquito control. *The American Journal of Tropical Medicine and Hygiene, 103*(3), 976–985.
12. Sánchez-León, S., Gil-Humanes, J., Ozuna, C. V., Giménez, M. J., Sousa, C., Voytas, D. F., & Barro, F. (2017). Low-gluten, nontransgenic wheat engineered with CRISPR/Cas9. *Plant Biotechnology Journal, 16*(4), 902–910. https://doi.org/10.1111/pbi.12837
13. Genetic Literacy Project. (2020). *United States: Therapeutic/stem cell*, https://crispr-gene-editing-regs-tracker.geneticliteracyproject.org/united-states-gene-therapy-stem-cells/
14. Regnault, H. (2022, January). La ménagerie impériale, *La Crise* n°42, CEIM UQÀM, https://www.ieim.uqam.ca/spip.php?page=article-ceim&id_article=13843
15. WHO. (2019). *WHO expert panel paves way for powerful international governance on human genome editing*. Press Release, https://www.who.int/en/news-room/detail/19-03-2019-who-expert-panel-paves-way-for-strong-international-governance-on-human-genome-editing
16. European Commission. (2021). *EC study on new genomic techniques*. Commission Staff Working Document SDW (2021), 92, https://ec.europa.eu/food/plants/genetically-modified-organisms/new-techniques-biotechnology/ec-study-new-genomic-techniques_en

Chapter 15
NGT and Animal Applications

Will second-generation biotechnologies face the same difficulties as first-generation techniques (see Chap. 10)? Or will they be able to propose new approaches to improve veterinary health and livestock breeding conditions? What animal applications are NGTs directed toward?

A few avenues are emerging from current research.

Laboratory Models

Like transgenic animals, animals are genetically edited by NGTs to understand the function of metabolic pathways and genes in several human diseases, inherited genetic diseases, and cancers.

Mice and rats, as well as dogs, pigs, and monkeys, are used for this purpose. For example, researchers at the *Yunnan Key Laboratory of Primate Biomedical Research* (China) are genetically modifying pigs and monkeys using CRISPR to create models to understand the mechanisms of muscular dystrophy, autism, cancer, and even resistance to HIV (AIDS virus).

Research in China is particularly developed on the animal models that are physiologically closest to humans, such as monkeys, because animal experiments on higher mammals are not culturally considered in the same way in Chinese society as in societies in Western countries (the United States, European Union). This Chinese research was able to develop from the training of Chinese researchers in the United States (Randall Prather's laboratory at the *University of Missouri* at Columbia in particular) who subsequently returned to China [1].

C. Regnault-Roger, *Biotech Challenges*, https://doi.org/10.1007/978-3-031-38237-6_15

Veterinary Medicine

In addition to this research on animal models to improve human health, veterinary medicine is not left behind, and research is developing in several directions: preventing and curing zoonoses[1] and epizootics[2], as well as pathologies affecting the performance of animal husbandry. Here are some examples.

Among the projects that stood out, work was undertaken to combat bovine tuberculosis, which is increasingly resistant to antibiotics, by *Shaanxi University* in China (College of Veterinary Medicine): this disease affects over 10 million animals worldwide. After complex manipulations on bovine fetal fibroblasts, Dr. Yong Zhang's team introduced the *MRAMP1* resistance gene into the cow's genome using CRISPR/Cas9. The researcher was pleased to have avoided *off-target* effects in the offspring of genetically edited animals [2].

Other research is being conducted to provide a solution to swine fever through a gene for resistance to this disease that is decimating entire herds in China and several other countries. In the United States, the *University of Missouri* is working with the British biotechnology company *Genus* to develop pigs resistant to the *porcine reproductive and respiratory syndrome virus* (PRRS). In partnership with the American company *Acceligen*, research is underway to develop pigs, but also cattle, to acquire resistance to FMDV (*foot-and-mouth disease virus*) [3].

In Japan, the *National Agriculture and Food Research Organization* (NARO) is working on CRISPR/Cas9 to combat the IARS (isoleucyl-tRNA synthetase) disease, named after the enzyme that codes for a protein that causes a perinatal calf weakness syndrome [4]. This disease is the result of a recessive mutation and causes high mortality in black Wagyu cattle, a meat breed with a high degree of marbling that is popular with Japanese meat lovers.

Animal Welfare

Research is also being conducted to improve animal welfare. For example, a Chinese team from a laboratory of the *Chinese Academy of Sciences* based in Beijing has been working with CRISPR to reduce the cold sensitivity of piglets, thus reducing their mortality rate in winter [5].

But the most spectacular research in terms of animal welfare concerns the obtaining of cows without horns. The dehorning of cattle is practiced in order to limit the accidental but frequent injuries of the animals and the cowherds. It is a particularly distressing operation for the animals that undergo it and the men who perform it. However, there are breeds of cows without horns: the Angus breed, highly appreciated for its meat, has a dominant hornless gene that is transmitted to its descendants.

---

[1] Zoonoses are diseases that can be transmitted from animals to humans and vice versa.

[2] An epizootic is an epidemic that affects animals of the same or different species in a given area.

This is the result of a voluntary selection process that takes place over a period of about 20 years [6]. In contrast, the Holstein breed, which is highly prized for its high milk productivity, has horns. The TALEN technique has made it possible to obtain hornless Holstein cows in one generation from genetically modified embryos that have integrated the hornless gene, whereas conventional genetic selection would have taken about 20 years. The prospects opened a significant improvement in animal welfare.

In a similar vein, work is being done by the company *Acceligen* in Minnesota to avoid chemically or surgically castrating pigs. This company is particularly active in using genome editing in animal welfare research, and its motto is "The animals that feed us deserve the best health possible." Its research concerns cattle, pigs, and fish [3].

Livestock Performance for Food Production

Genome-editing techniques are also being used to improve the performance of livestock.

TALEN and ZFN techniques have been used to increase muscle mass in domestic animals. Cows, sheep, and pigs have been inhibited for the gene responsible to produce myostatin, a factor that limits the growth of muscle tissue [7]. Research is being conducted by the *Chinese Academy of Agricultural Sciences* and the National University in Yangling (Xianyang) (*Northwest A&F University*) and *Yanbian University* in South Korea to obtain goats and pigs with increased muscle mass and decreased fat mass [8]. Projects in the early R&D stages are investigating the reduction of sterility to improve reproductive performance and the modification of behavioral traits in the herd. The production of milk with hypoallergenic properties is also being investigated.

In fish farming, requests for authorization were presented to the National Technical Commissions of Biosafety of the Brazilian authorities (CTNBio(*Comissão Técnica Nacional de Biossegurança*)) and Argentina (CONABIA (*Comisión Nacional Asesora de Biotecnología Agropecuaria*)) for tilapia farms with accelerated growth (+16%) and increased yields (+70%) with a feed conversion ratio of +14% [9]. This research, led by the American company *Precigen* (*ex-Intrexon*), has been developed using the technology developed by *AquaBounty Technologies*, which markets a transgenic salmon with accelerated growth (see Chap. 10). They are at an advanced stage, with significant results suggesting that they will soon be marketed.

Of the 63 R&D projects identified for animal applications for commercialization, two thirds concern domestic animals and fish farms (43 projects), half of which are at an advanced stage of development. The most widely used technique is CRISPR (75%).

What About Europe?

European research, through its FP7 and H2020 framework programs, has funded 19 basic research projects in the animal field from 2007 to 2020, for about 90 million euros [10]. They focus on the understanding of genetic mechanisms in immunity and adaptive evolution, cell differentiation, and interspecific animal-bioaggressor relationships.

In France, the position of the Ethics Committee common to three French research organizations INRAE, CIRAD, and IFREMER[3] on the interest of conducting research with NGTs is to be noticed. In December 2019, this committee pronounced on the use of new genome-editing techniques, including CRISPR/Cas9, for animal applications [11]. It emphasized the interest of agronomic applications of these techniques as a tool to advance knowledge but also for finalized research purposes.

It identified the following axes:

- Improvement of animal husbandry through research to improve animal welfare, enhance the ability of animals to make the most of the food they receive, and strengthen their resistance to disease and their adaptation to their environment.
- Reduction of the environmental impact of livestock farming.
- Control of pests, parasites, or invasive species that threaten native species and biodiversity, including further work on genetic forcing.

However, the Ethics Committee recommends caution in the use of genome-editing techniques applied to animals by insisting on the notion of boundaries to be "assigned."

Conclusion

There is a consensus on the usefulness of NGTs both for fundamental and applied research in veterinary health and animal welfare and for farm performance.

The President of the *French Veterinary Academy* for year 2021, Jean-Pierre Jégou, recently stated [12] that:

> "The opinion of the French Veterinary Academy is clearly in favour of the development of research projects using modern genome engineering technologies." He is "convinced that some of the applications will be able to contribute to meet urgent global challenges such as the fight against zoonotic panzootics[4]."

---

[3] INRAE, National Research Institute for Agriculture, Food and Environment; CIRAD, Center for International Cooperation in Agricultural Research for Development; IFREMER, French Research Institute for Exploitation of the Sea.

[4] Pandemics for animals.

He also points out that:

> Several North American countries, the People's Republic of China and the United Kingdom have already produced pigs in 2017 that are insensitive to PRRS (Porcine Respiratory Distress Syndrome) virus, with an estimated annual induced global loss of $2.5 billion.

In conclusion, he deplores the fact that genomic editing in animals is "ignored or obscured" at both French and European levels.

Thus, for the French Veterinary Academy, the research on NGTs for animals applications should be "encouraged at all levels and adequately financed, otherwise it will result in a delay that will be detrimental to Europe," for therein lies the challenge: that of agri-food independence in the animal production sector.

Bibliography

1. Cohen, J. (2019, June 31). China's CRISPR push in animals promises better meat, novel therapies, and pig organs for people. *Science*, https://www.sciencemag.org/news/2019/07/china-s-crispr-push-animals-promises-better-meat-novel-therapies-and-pig-organs-people
2. Gao, Y., Wu, H., Wang, Y., Liu, X., Chen, L., Cui, C., Liu, X., Zhang, J., & Zhang, Y. (2017). Single Cas9 nickase induced generation of *NRAMP1* knocking cattle with reduced off-target effects. *Genome Biology, 18*, 13. https://doi.org/10.1186/s13059-016-1144-4
3. Acceligen. (2021). *Traits for improved animal health and well-being*, https://www.acceligen.com/
4. Ikeda, M., Matsuyama, S., Akagi, S., Ohkoshi, K., Nakamura, S., Minabe, S., Kimura, K., & Hosoe, M. (2017). Correction of a disease mutation using CRISPR/Cas9-assisted genome editing in Japanese black cattle. *Scientific Reports, 7*, 17827. https://doi.org/10.1038/s41598-017-17968-w
5. Zheng, Q., Lin, J., Huang, J., Zhang, H., Zhang, R., Zhang, X., Cao, C., Hambly, C., Qin, G., Yao, J., Song, R., Jia, Q., Wang, X., Li, Y., Zhang, N., Piao, Z., Ye, R., Speakman, J. R., Wang, H., Zhou, Q., Wang, Y., Jin, W., & Zhao, J. (2017). UCP1 decreases fat deposition in pigs. *Proceedings of the National Academy of Sciences, 114*(45), E9474–E9482. https://doi.org/10.1073/pnas.1707853114
6. Houdebine, L. -M. (2017). Les vaches sans cornes. *Science and Pseudoscience, 321*, 12.
7. Houdebine, L. -M. (2018). Les nouveaux outils des biotechnologies animales. In C. Regnault-Roger, L. -M. Houdebine, & A. Ricroch (dir), *Au-delà des OGM* Presses des Mines (pp. 67–92).
8. Global Gene Editing, Regulation Tracker. (2021). *China animals*, https://crispr-gene-editing-regs-tracker.geneticliteracyproject.org/china-animals/
9. The Fish Site. (2018, December 18). *Gene edited tilapia secure GMO exemption*, https://thefishsite.com/articles/gene-edited-tilapia-secures-gmo-exemption
10. European Commission. (2021). *EC study on new genomic techniques*, Commission Staff Working Document SDW (2021), 92, https://ec.europa.eu/food/plants/genetically-modified-organisms/new-techniques-biotechnology/ec-study-new-genomic-techniques_en
11. INRAE. (2019, December 13). *avis du comité d'éthique* INRA-CIRAD-IFREMER *opinion of the* INRA-CIRAD-IFREMER *ethics committee*, pdf 83208KB and Nicole Ladet: *La modification génétique des animaux à l'épreuve de l'édition du génome*, https://www.inrae.fr/actualites/modification-genetique-animaux-lepreuve-ledition-du-genome
12. Jégou, J. –P. (2021, June 4). L'Europe doit évoluer sur l'édition génomique animale. Europe must evolve on animal genome editing. Interview. *European scientist*, https://geneticliteracyproject.org/2021/08/09/crop-chemophobia-ii-when-activist-journalists-twist-science-in-support-of-ideology/

Chapter 16
Second-Generation Agricultural Plant Biotechnologies: State of the Art

Plant production is still today the leading sector in the development of biotechnologies with the boom in the cultivation of transgenic plants (see Chap. 12). It is therefore quite natural that research and development (R&D) projects have multiplied in this field, as the editing of the plant genome is technically easier than that of the animal genome.

As has been pointed out, the latest genome-editing techniques are more precise, faster, and less expensive than transgenesis or random mutagenesis, and as a result, they were very quickly integrated into the toolbox of seed breeders to optimize the varietal improvement of cultivated plants. This is why the scientific community spontaneously called them *new plant-breeding techniques* (NPBT), before their name was broadened by first removing the reference to "plant" (*new breeding techniques* (NBT)) and then generalizing the object (modification of the genome to all living organisms) with the name NGT (*new genomic techniques*) according to the name used by the European Commission (cf. Chap. 4).

With such precision tools, the stages of conventional plant breeding for varietal improvement have been considerably lightened, and research in the plant sector is developing intensively. As we have already indicated (see Chap. 13), this sector accounts for two-thirds (66%) of current R&D projects, with 210 projects at an advanced stage. However, few products resulting from NGTs are already on the market.

Dynamism in Plant Breeding Research

A recent review of the scientific literature identified publications devoted to *genome-editing* applications in agricultural and model plants between January 1996 and May 2018 [1]. As for other sectors (medical and animal applications), the CRISPR technique discovered in 2012 has mobilized researchers the most: 1032 articles are

© The Author(s), under exclusive license to Springer Nature Switzerland AG 2023
C. Regnault-Roger, *Biotech Challenges*,
https://doi.org/10.1007/978-3-031-38237-6_16

devoted to it out of the 1328 studies identified. A second study based on a census of R&D projects until 2021 confirms the pre-eminence of the CRISPR technique as the preferred tool for breeders compared to other techniques: CRISPR, 70.8%; TALEN, 7.4%; ODM, 8.4%; and meganucleases, 1.4% [2].

Twenty-eight species of cultivated plants are involved: the importance of rice is noteworthy, which is not surprising when one considers that the Asian continent alone has driven more than 50% of these research publications. The second most explored plant is the model plant *Arabidopsis thaliana*, followed by tobacco. Model organisms are studied by the entire international scientific community and provide benchmarks for scientific advances. The species following are food crops: tomato, wheat, corn, soybean, and potato (Fig. 16.1). Many other plants have been identified since then: ornamental plants, fruit trees, sugar plants (beet and cane), oil plants, tubers, forage plants, etc. [2]. Virtually, all crops are the subject of NGT's breeding research.

The aim of these genomic modifications is primarily to improve agronomic traits (technical itineraries, growth, yield), food quality (human and animal nutrition), tolerance to biotic stresses (insect pests, diseases, viruses, etc.) and abiotic stresses (drought, nitrogen deficiency, UV radiation, etc.), tolerance to herbicides (weed control), and industrial valorization.

Most transformation techniques are minor modifications of type SDN1 and point of mutation or basic edition (92%), and for the more consequent modifications of the genome, 3% are classified as SDN2 and 5% as SDN3.

Globalized Research

Research was conducted in 68 countries, with two countries being particularly active: China (599 publications or 40%) followed by the United States (487 publications or 33%).

Among the laboratories and companies that are at the forefront of this research, there is a strong investment by Chinese organizations (*Chinese Academy of Sciences* and the agricultural academies of various provinces as well as Chinese universities) for studies not only on rice, for which they are carrying out intense activities, but also on rapeseed, cotton, tomato, and wheat. China is a leader in agricultural applications. In fact, China has by far the most patents for agricultural applications (see Chap. 13).

Many American universities (California, Minnesota, Florida, Kansas State, Penn State), some universities in Japan (Tamagawa, Tsukuba, Tokushima) and in Europe (University of Kiel in Germany, Universities of Paris-Saclay and Montpellier in France), and well-known research centers such as the *Weizmann Institute of Science* in Israel, the *Max Planck Institute* in Germany, and the *Instituto de Agricultura Sostenible* of the CSIC (*Consejo Superior de Investigaciones* Científicas) in Spain

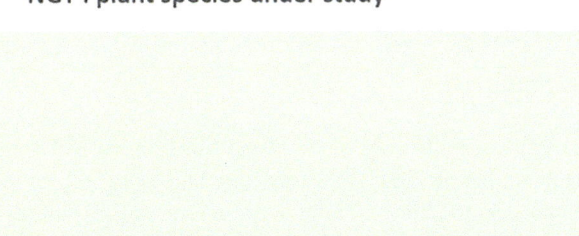

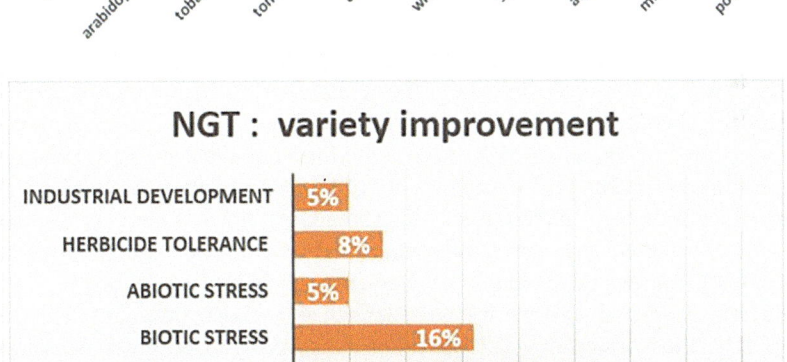

Fig. 16.1 Applications of NGTs – most studied plants between 1996 and 2018 and object of varietal improvements. (Based on data from Modrzejewski et al. [1])

are very involved in this research. Private sector companies (DuPont Pioneer, Syngenta Seeds, Cellectis Plant Sciences which became Calyxt in 2010, Dow AgroSciences) are not to be outdone.

What research is being developed today in the various countries and what stage are the projects at? In order to understand the general dynamism of these developments by NGT, we will describe the situation of the most committed countries of each continent, based on examples that seem significant to us, before outlining the general perspectives.

Situation by Continent

North America

United States

This large country is very much at the forefront of NGTs development, with over 763 patents filed on CRISPR technology but "only" 61 for agricultural applications (in comparison to 259 for China out of a total of 847 patents and 18 for Europe out of a total of 188) over the period 2013–2018 [3]. This US investment in NGTs is not new, which is why techniques that predate the discovery of CRISPR, such as TALEN or ZNF, have been used extensively in agricultural applications.

The American company *Calyxt* has developed several research projects with the TALEN technique to improve the quality of alfalfa and the resistance of wheat to mildew. It has also edited soybeans to produce soybean oil without transfatty acids and with less saturated fatty acids, which improves its nutritional quality (see Chap. 14). For field crops, it has developed mildew-resistant wheat varieties. This pioneering company is also very involved in the research of *Innate® potato* varieties (cf. Chap. 10). These potatoes marketed by Simplot Plant Sciences have been the subject of campaigns by anti-GMO activists who have convinced the world's largest manufacturer of frozen potato products, *McCain Foods*, to abandon the use of these potatoes. However, given their processing technology, the new *Innate®* potato varieties are considered to be genetically edited and have been deregulated (see Chap. 10). Will they have a better chance of conquering the American market?

Several American universities have launched research programs, some with TALEN but the most recent with CRISPR on many species. Here are some examples:

- The *Penn State University* has developed a button mushroom (*Agaricus bisporus*) genetically edited by CRISPR so that it does not turn brown. It was approved for commercialization by the FDA back in 2016 without being subject to GMO regulations.
- *Rutgers University* develops blight-resistant grapevine using CRISPR. The *University of California Riverside* has CRISPR-edited a mini-tomato so that it can be grown in limited spaces (*indoor* growing, at home, in a vertical farm or in the international space station!).
- The MIT (*Massachusetts Institute of Technology*) is developing the creation of nitrogen-fixing cereals (corn, wheat, rice), which will revolutionize their field crops since it will no longer be necessary to apply nitrogen fertilizers [4].
- The *University of Iowa* is using TALEN to develop a rice resistant to the pathogenic bacterium *Xanthomonas oryzae* pv. *oryzae.*
- The *University of Minnesota* is using CRISPR to create halophilic, drought-tolerant soybeans.

Other organizations or companies have developed genome-editing products: *Yield10 Bioscience* (*Camelina* with increased omega-3 fatty acid content) and *DuPont* (drought-resistant corn, waxy corn). Several of these products have been "deregulated" (exempted from GMO regulation) and authorized by the USDA (*United States Department of Agriculture*). However, the marketing of these products depends on the logistics of the companies that market them. It can take several years from approval to commercialization.

Canada

Research is being developed by Ruimin Rao's team at *Agriculture and Agri-Food Canada* to CRISPR-edit alfalfa (*Medicago sativa*) in order to better understand the possible improvement pathways for this plant [5].

Among the crops that have received approval over the past decade are apple trees whose apples are marketed under the *Arctic®* label by *Okanagan Specialty Fruits Inc.*, a small British Columbia company (acquired by US company *Intrexon* in 2015). Apple growers in the Okanagan region decided to add value to their local production by creating fruit that can be stored once cut, without browning in ready-to-use packages. Apples of the *Granny Smith* and *Golden Delicious* varieties, followed by the *Fuji* variety, were modified to resist browning due to the oxidation of certain constituents (polyphenols) that these fruits naturally contain, and that occurs when the fruit is shocked or cut. As for the *Innate®* potato, the genetic modification technique for *Artic®* apples is complex and uses RNAi. *Artic®* varieties have been deregulated in the United States and are not considered GMOs [6]. They have also been approved by *Health Canada* and are approved for sale in that country.

Latin America

Argentina

The Argentine Institute of Agricultural Technology (INTA (*Instituto nacional de tecnología agropecuaria*)) is piloting CRISPR projects for high-quality alfalfa and non-browning potatoes (without polyphenol oxidase) with field trials started in 2020 and a cotton plant resistant to the weevil *Anthonomus grandis*.

Brazil

The *University of Viçosa* is developing spicy tomatoes rich in capsaicinoids and, for research purposes, tomatoes high in lycopene (an antioxidant that protects against free radicals). The Brazilian company *Tropical Melhoramento & Genética S/A* has launched a collaboration with the Israeli company *Evogene* to develop a soybean resistant to the rootworm *Heterodera glycines*, a major nematode pest of the plant. Other projects have been submitted to the Brazilian National Technical Commission for Biosafety (*Comissão Técnica Nacional de Biossegurança* (CNTBio)), including the genetic editing of fungi to produce bioethanol.

Uruguay

The National Institute of Agricultural Research (INIA (*Instituto Nacional de Investigación Agropecuaria*)) and the University of the Republic (*UdelaR*) are developing projects to breed glyphosate-resistant soybean varieties (which have already been bred by transgenesis, so the aim is to shorten the time needed to breed a variety by using a technique that is easier to implement) and also to improve the nutritional profile of soybeans by reducing lectins. This technique is also used to increase the lycopene content of tomato and mandarin varieties.

Chile

The projects aim at the nutritional improvement of soybean and *Camelina* edible oils by modifying their unsaturated fatty acid contents. Studies have also been carried out on corn that develops tolerance to excess water.

Costa Rica

In Central America, the University of Costa Rica and CENIBiot (*Centro Nacional de Innovaciones Biotecnológicas*) are developing a project to breed drought-resistant rice to respond to climate change.

Asia-Pacific Zone

China

It is not surprising that Chinese research is focused on rice, a native crop and essential food consumed by the population. Research is being done to increase yields by 30% (agronomic interest), to increase fiber content (nutritional interest), and to modify the fragrance of rice to give it a jasmine scent that appeals to the public (commercial interest). The *Chinese Academy of Sciences* has also developed high-yielding varieties of wheat, corn, and soybeans, as well as HT varieties and those better adapted to high heat. Other studies are being conducted on kiwi, poplar, grapes, and tomatoes.

Chinese laboratories conducted early research with the TALEN technique, but benefiting from the biotechnological expertise of the international company Syngenta (see below) they are now favoring the CRISPR technique. They also have many managers trained at North American, Australian, and European universities who have maintained relationships with their former universities and have returned home. Many exchanges are taking place between Chinese and foreign laboratories in a favorable context since the Chinese authorities are encouraging varietal improvement by NGTs [7].

Japan

Research is being conducted by the National Agriculture and Food Research Organization (NARO (*Nōgyō Shokuhin Sangyō Gijutsu Sōgō Kenkyū Kikō/Japanese National Research and Development Agencies*)), whose research center is located in Tsukuba, to obtain high-yielding rice (rice also being a staple of Japanese cuisine) but also wheat that is resistant to excess water. NARO is developing with the Universities of Tsukuba and Yokohama research to modify the purple color of the emblematic flowers of Japan, the *Japanese morning glory* (*Ipomoea nil* L.), and obtain white flowers by CRISPR: a single gene is involved in this change of color which was obtained in 2017 [8].

More generally, research is very active in all specialized university laboratories. For example, researchers at the University of Tokyo have created new lines of rice and canola by genome editing using mitochondrial DNA, using a new technique called "mitoTALEN" in which they hold out hopes for increasing agricultural biodiversity and promoting food security [9]. Researchers at *Tokushima University* have developed seedless tomatoes for research purposes only.

Beyond the laboratories, Japan has authorized the marketing of the CRISPR/Cas genetically edited *Sicilian Rouge High GABA* tomato in 2021 (see Chap. 14).

India

The *Indian Council of Agricultural Research (ICAR)-National Institute For Plant Biotechnology* has been using CRISPR since 2019 to develop halophilic (salt-tolerant) rice, and researchers at the *National Agri-Food Biotechnology Institute (NABI)* want to create a vitamin A-rich biofortified banana, similar to golden rice, to combat this avitaminosis that plagues developing countries.

Australia and New Zealand

The Australian CSIRO (*Commonwealth Scientific and Industrial Research Organisation*) and its laboratories have developed research programs with different techniques:

- RNAi to obtain high-yielding wheat with increased quality and tolerance to environmental stresses, heat, cold, and drought, as well as barley resistant to BYDV (*barley yellow dwarf virus*) and a cotton plant modified so that the oil extracted from the cakes has less transfatty acids, which improves its nutritional profile.
- TALEN and CRISPR for wheat rust resistance.

Laboratories at the Universities of Queensland, Sydney, and Murdoch are using CRISPR to improve the drought tolerance of canola, the dietary profile of sorghum with high-protein content, gluten-free potatoes, and glutinous rice.

New Zealand, which is legally very reticent about biotech, is developing few research programs in the field of NGTs. Moreover, experimental field trials are conducted outside the country. They are hosted in the United States, such as the work of the *New-Zealand AgResearch Center*, which aims to obtain a drought-tolerant forage plant, the ryegrass (genus *Lolium*), in order to cope with climate change and, those devoted to developing grass with reduced methane emissions, a greenhouse gas. Many voices are being raised to change this situation [10].

Europe and Mediterranean Basin

The United Kingdom

The well-know *John Innes Centre* and *Rothamsted Centre* are developing programs on a number of plants, including *Camelina* to obtain a high oleic acid content for food purposes and beets biofortified with L-Dopa to combat Parkinson's disease. Work is also being done to improve nitrogen nutrition in barley by CRISPR, which can grow faster. The Brexit, which has freed the United Kingdom from EU regulations, is encouraging the biotech research.

Sweden

The Swedish University of Agricultural Sciences (SLU (*Sveriges Lantbruksuniversitet*)) in Alnap is developing research to increase amylose, a constituent of potato starch, for industrial applications (paper, textiles, glues, etc.).

The Netherlands

Wageningen University (*Wageningen Universiteit en Researchcentrum* (WUR)) is conducting research with CRISPR on gluten-free wheat and starch-enhanced potatoes for nonfood use.

Belgium

A project for a banana resistant to a pathogenic fungus has been abandoned after the Court of Justice of the European Union (CJEU) ruled on July 25, 2018, that genome-editing products are GMOs. The Flemish Research Center VIB (*Vlaams Instituut voor Biotechnologie*) at Ghent University is continuing its work with CRISPR to create new climate-stress-adapted corn varieties for research purposes only. Field trials are underway [11].

France

A large-scale research project Genius [12] led by INRAE is using genome-editing techniques, including TALEN and CRISPR/Cas9, on nine crop species (wheat, maize, rice, tomato, potato, rapeseed, poplar, apple, rose) and three model plants (*Brachypodium, Physcomitrella, Arabidopsis*) in order to establish a proof of concept[1] concerning various traits that these species could acquire such as resistance to diseases, tolerance to salinity, flowering precocity, plant architecture, or reproduction. The objective is to propose improvements to these plants for agricultural sustainability. Among the research projects, work on the resistance of tomatoes to pathogenic potyviruses and the reduction of amylose in potatoes in order to obtain a starch sought after by industry (food, cosmetics, paper, etc.) are well under way.

The aim of the Genius project is to provide "genome editing technology to French public and private research teams for fundamental research in plant biology in cultivated species" and to structure a community "around a major challenge, the implementation of genome editing in plant biology in France."

In addition to various INRAE laboratories and the University of Lyon 3, this project has already brought together the private companies Biogemma, Germicopa, Delbard, and Vilmorin. It also "irrigates" (according to the expression used by its leaders) 35 French, European, and foreign partners (the United States, Japan, Mexico, Russia, China, Argentina) with the results of its research.

Alongside this program, other partnerships have also been initiated to provide European seed companies with access to US patent-protected CRISPR-based genome-editing tools for agricultural applications. The agreement signed in December 2019 between Corteva Agriscience, MIT and Harvard (Center for Genomic and Biomedical Research), and the company Vilmorin illustrates the approach [13].

Russia

An ambitious federal program has been developing since April 2019. It focuses on genome editing of four priority crops, barley, sugar beet, wheat, and potato, of which Russia is one of the world's main producers (leader for barley). The *Russian Academy of Sciences* (RAS) in Moscow is already developing research to make potatoes and sugar beets more resistant to various diseases. The *Vavilov Institute* in Saint Petersburg, a historical research center created in 1894, today called "All-Russian Institute of Plant Genetic Resources," and the Federal Institute of Cytology and Genetics of RSA, Siberian branch (Novosibirsk), undertook research on varietal improvement of barley and wheat by NGT [14].

---

[1] The purpose of a proof of concept is to show the feasibility of a process or an innovation.

Israel

A variety of research is being developed by historical biotechnology institutes such as the *ARO Volcani Centre* (*Agricultural Research Organization*) founded in 1921 in the Tel Aviv area, as well as *Yissum*, "the bridge between research at Hebrew University and the world." It is the technology transfer company of the *Hebrew University of Jerusalem* (founded in 1964), whose objective is "to translate promising results from university research laboratories into commercial applications" [15]. They are developing projects for tomatoes resistant to the Egyptian broomrape (*Phelipanche aegyptiaca*), a parasitic plant on which conventional solutions (herbicides or mechanical removal) are ineffective [16], or for making cucumbers resistant to several viruses (the potyviruses ZYMV (*Zucchini yellow mosaic virus*) and PRSMV (*Papaya ringspot mosaic virus*) and the ipovirus CVYV (*Cucumber vein yellowing virus*)). Close collaborations between the Israeli horticultural company *Danziger* and the American company *Precision BioSciences* are being conducted to modify the color of solanaceous plants of the genus *Petunia* and of an ornamental plant, jasmine tobacco (*Nicotiana alata*). Research to biofortify food crops (potato, wheat, corn, canola, tomato, and cucumber) by genome editing is very active in Israel.

Egypt

The laboratory of Prof. Naglaa Abdallah at Cairo University (Department of Genetics, Faculty of Agriculture) is conducting research with CRISPR to develop drought-tolerant wheat varieties [17].

African Continent

South Africa has historically been a proponent of transgenic crops and has been growing them since 1996, hosting numerous pre-cultivation trials over the past decade: it is a pioneer country. It has been joined by other countries in recent years. However, few of these African countries, which have now agreed to grow biotech plants on their territories, have the capacity to develop research using NGTs. Nevertheless, a few university laboratories or research centers specialized in tropical agriculture are studying the improvement of plants that are very important for African agriculture. These laboratories are mostly located in Eastern and Southern Africa (Box 16.1).

This African university research is supported by international organizations such as the *International Institute of Tropical Agriculture* (IITA) and the *International Maize and Wheat Improvement Center* (CIMMYT), American universities (*Penn State University* and the Innovative Genomics Institute (IGI) at the *University of California* at Berkeley and San Francisco), and Australian universities (*University of Queensland*). They are conducting studies in collaboration with local organizations to propose varietal improvement solutions adapted to African agriculture.

Cassava (*Manihot esculenta*) is widely cultivated in rural areas of tropical and equatorial Africa. Prolonged consumption of its tuber, which is rich in starch but also in cyanogenic glucosides, causes Konzo disease, a neuromotor disorder that can degenerate into paralysis, during periods of famine and drought. Indeed, cyanogenic glucosides can be transformed into cyanide if, due to the lack of water, the cassava is badly prepared. The California-based IGI is conducting research with CRISPR to disable two genes *CYP79D1* and *CYP79D2* involved in the synthesis of cyanogenic glycosides in order to reduce their content in cassava. Other research is also being carried out to CRISPR-edit a cassava resistant to the *cassava brown streak* virus (CBSV). Field trials are being conducted in Uganda and Kenya by the *Donald Danforth Plant Science Center* in St. Louis, Missouri.

IITA is also conducting CRISPR research to develop banana plants that are drought-tolerant or drought-resistant to diseases caused by *Erwinia* bacteria in Nigeria and Kenya (Dr. Leena Tripathi's laboratory in Nairobi).

Box 16.1: Main African University Laboratories Developing Plant Biotechnology Research Using Genome-Editing Techniques (NGT) According to Karembu [17]

South Africa: The South African laboratory led by Prof. Chrissie Rey of the Witwatersrand University in Johannesburg (*Wits University, School of Molecular and Cell Biology*). This renowned scientist is conducting research to find a solution to cassava mosaic caused by the *South African cassava mosaic virus* (SACMV) by studying the role of genes involved in the disease to silence them.

Kenya: *Kenyatta University* in Nairobi (Prof. Steven Runo) is working on sorghum to develop varieties that are CRISPR-/Cas9-resistant to striga, a plant pest that is prevalent in Kenya. Dr. James Kamau Karanja's laboratory, which has a dual affiliation with KALRO/NARL (*Kenya Agricultural and Livestock Research Organization* and *National Agricultural Research Laboratories*) in Kabete, is looking for genetic solutions to maize lethal necrosis, an emerging African disease.

Ethiopia: Research is being conducted with CRISPR on Ethiopian mustard (*Brassica carinata*) to improve the food quality of its oil by the team of Prof. Teklehaimanot Haileselassie Teklu of the *Institute of Biotechnology* at Addis Ababa University.

Uganda: Dr. John Odipio is conducting research at the *National Crops Resources Research Institute* (NaCRRI) on the Namulonge campus to develop high-yielding, stress-resistant cassava from varieties favored by Ugandan farmers. He is also studying rice resistance to RYMV (*rice yellow mottle virus*) and maize resistance to lethal necrosis caused by two potyviruses.

A final example is cocoa farming in Ghana, where production is facing a disease caused by CSSV (*cacao swollen shoot virus*) and climate change. The Ghana *Environmental Protection Agency* and the *Cocoa Research* Institute *of Ghana* predict that cocoa farming could disappear from the country as early as 2080 if these scourges are not addressed. Research is being conducted with *Penn State University* (USA) to provide an answer to save the African cocoa crop with genetically edited cocoa trees by CRISPR [17, 18]. This echoes the development of transgenic papaya trees (called *Rainbow papaya* by the *Hawaii Food Industry Association* because they gave Hawaiian farmers "a second chance") a few years ago (first-generation biotechnology), which saved this food crop that was about to disappear from Hawaii Island.

This global overview highlights a real abundance of research in varietal improvement. It is important to underline that these developments are the result of numerous international collaborations and intense exchanges between researchers. Research in this sector is globalized.

Agricultural NGTs in Perspective

A Chinese Desire to Match the United States

It is not surprising that the bulk of agricultural applications of NGTs are currently in China and the United States, as these two countries have the most patents in the field, with China far ahead of the United States as noted earlier (see Chap. 13). However, the size of the US market is ten times larger than that of the Chinese market, and of the 802 biotech companies listed on the stock market with a capitalization of USD 1060 billion, 408 are American, representing 52% of the global valuation, while Chinese companies, 37 in number, account for only 10% of the world's market capitalization [19].

However, the biotechnology sector has been considered a priority by the Chinese government for many years, and the start of these efforts dates to 1978 [20]. It is benefiting from the growing importance of state funding. The acquisition of the Syngenta group for USD 43 billion (EUR 39 billion) in 2017 by the state-owned *China National Chemical Corporation* (see Fig. 16.1 in Chap. 12) gives China access not only to the European market but to the worldwide market. The merger allows China to use Syngenta's protected seeds, as Aifang Ma notes. It also allows China to benefit from "a large R&D team working with CRISPR/Cas," says Agnès Ricroch, a member of the French Academy of Agriculture, who adds that "the *Syngenta Beijing Innovation Center* works in close collaboration with the 'sister R&D' in Durham, North Carolina, in the United States" [21]. This example illustrates the desire of the Chinese political authorities to encourage the strategy of mergers and acquisitions in the biotechnology sector in order to compete on a par with the United States.

But this global openness needs to be accompanied by common rules. For example, trade relations between the United States and China in the field of agricultural biotechnologies, under the "Phase One Agreement" signed by the two countries in

2020, show that China must be "more transparent and predictable" in its GMO authorization procedures, according to the US administration in November 2022 [22].

R&D in Public/Private Partnership for Marketing

If many plants are the subject of research, around 60 species, the most studied plants are the three plants that biotechnologists place "in the Pantheon of model plants" according to the classification established by Alain Toppan [23]: rice, *Arabidopsis thaliana*, and tobacco (two of them are economically major crops in China), followed by tomato, maize, wheat, soybean, and potato [24].

The 409 R&D projects are expected to be completed around 2030 (see Chap. 13). They concern yield increases, crop pest control, resistance to abiotic stresses (heat, drought, salinity), and development of biofortified crops with a higher content of certain nutrients or through the reduction of anti-nutritional compounds (allergens, cyanogenic glucosides that are precursors of acrylamide) or gluten. This list is not very different from the publications cited above (Fig. 16.1), each of which precedes the other.

There are 15 research projects that are in the pre-commercialization phase: tolerance to one or more herbicides, fungal resistance, modification of starch nutritional composition, and non-browning by enzymatic inhibition of apples or of the button mushroom (*Agaricus bisporus*). Some products have already received authorization from the competent authorities to be placed on the market, but few products derived from NGT have reached the marketing stage (see Chap. 14).

It is remarkable that the companies developing these new products are backed by public research laboratories. For example, in the United States, the American company Calyxt, a subsidiary of the French company Cellectis, has signed an agreement with the University of Minnesota for the commercial use of the TALEN technology [25], and in Japan, the *Sicilian Rouge High GABA* tomato is marketed by the company Sanatech, a technology transfer company backed by the University of Tsukuba. There is a very active involvement of academic research in R&D projects, while private sector companies are very present in the pre-commercialization and commercialization phases of second-generation biotechnology products. The NGT products currently marketed underline the need for public/private sector complementarity to reach the market.

In Europe

In terms of European research, the projects concern yields, resistance to biotic stresses (diseases) and abiotic stresses (climate change), and the production of metabolites of pharmaceutical interest or food-health products (fibers, fatty acids). The plants studied are wheat, vegetables, chicory, fruits, and also poplars.

Table 16.1 Species and traits studied by European seed companies (from the French Union of Seed Companies (UFS) [26, 28])

Species worked	Main characteristics studied
Field crops (cereals, corn, oilseeds, forage Specialty crops (vegetables, ornamentals, medicinal) Diversification species (legumes, hemp, dandelion, stevia, etc.)	Agronomic value (25%) Disease and pest tolerance (23%) Nutritional quality (18%) Abiotic stress tolerance (15%) Industrial applications (9%) Herbicide tolerance (5%)

Since 2007, 78 projects have been financed to the amount of 271 million euros by the European framework programs FP7 (2007–2014) and H2020 (2014–2020). Germany is the country that is developing the largest number of research projects [26] and is filing twice as many patents in the plant biotechnology sector in 2020 as France (see Chap. 13).

A survey conducted in the first half of 2020 by *Euroseeds*, the European Federation of Seed Companies, among its member companies on the use of NGTs, provides an overview of the use of these techniques by companies that are highly mobilized on varietal improvement and the marketing of new varieties. The survey was conducted among 62 companies representing 92% of the European breeders who are members of *Euroseeds*.

The French Union of Seed Companies (UFS) echoed this during a public hearing conducted in March 2021 by the OPECST [26, 28]. It indicated that all large companies (sales of more than €450 million) use NGTs, medium-sized companies (sales of between €50 million and €450 million) 85%, and small companies (sales of less than €50 million) 50%. As elsewhere in the world, a considerable number of species are processed by these companies, and the traits studied are similar in nature (Table 16.1).

This survey shows that NGT technology has become essential for European entrepreneurial innovation in the seed sector, which is known to be an important economic sector in France, the world leader in field crop seed exports ahead of the United States and the Netherlands in 2019 [27].

Appendix: Microorganisms

To complete this overview, it is necessary to mention genetically engineered microorganisms about which little information is available [2]. In agriculture, they can be used as soil biofertilizers (a commercialized application) or as biocontrol agents in crop protection. They are also mainly used as "biofactories" to synthesize compounds of pharmaceutical or cosmetic interest, enzymes that can be used in various nonfood (textiles, detergents, biofuels, etc.) as well as food (bakery, starch industry) industries. European funding over the last 13 years (2007–2020) has been granted to the sum of 135.5 million euros for 47 projects concerning microorganisms, some of which have agricultural implications [26].

Conclusion

These numerous applications of second-generation biotechnologies in plants highlight the wide range of these innovations. The conclusion is clear: in plant innovation as well as in animal innovation and human health, genome-editing techniques have become indispensable – more precise and less expensive, they open infinite horizons of innovation in biological sciences and in the life sciences industry.

Bibliography

1. Modrzejewski, D., Hartung, F., Sprink, T., Krause, D., Kohl, C., & Wilhelm, R. (2019). What is the available evidence for the range of applications of genome-editing as a new tool for plant trait modification and the potential occurrence of associated off-target effects: A systematic map. *Environmental Evidence, 8*, 27. https://doi.org/10.1186/s13750-019-0171-5
2. Parisi, C., & Rodriguez-Cerzo, E. (2021). *Current and future market applications of new genomic techniques*. Joint Research Centre (EUR 30589 EN), p. 49.
3. Martin-Laffon, J., Kuntz, M., & Ricroch, A. (2019). Worldwide CRISPR patent landscape shows strong geographical biases. *Nature Biotechnology, 37*, 601–621.
4. Miller, L., Latif, A., Jameel Water and Food Systems Lab. (2020, January 20). Making real a biotechnology dream: nitrogen-fixing cereal crops. *MIT News*, https://news.mit.edu/2020/making-real-biotechnology-dream-nitrogen-fixing-cereal-crops-0110
5. ISAAA Crop Biotech Update. (2018). CRISPR/Cas9 Found Applicable to Alfalfa, https://www.isaaa.org/kc/cropbiotechupdate/article/default.asp?ID=16265
6. Xu, K. (2015, April 29). Why arctic apples were approved By USDA. *Growing produce*, https://www.growingproduce.com/fruits/apples-pears/why-arctic-apples-were-approved-by-usda
7. Cohen, J. (2019, July 29). To feed its 1.4 billion, China bets big on genome editing of crops. *Science*, https://www.sciencemag.org/news/2019/07/feed-its-14-billion-china-bets-big-genome-editing-crops
8. Genetic Literacy Project. (2017, September 8). *In world-first, Japanese scientists use CRISPR to change flower*, https://geneticliteracyproject.org/2017/09/08/world-first-japanese-scientists-crispr-change-flower-color/
9. University of Tokyo. (2019, July 8). Researchers can finally modify plant mitochondrial DNA. *Nature Plants*. https://doi.org/10.1038/s41477-019-0459-z., https://www.eurekalert.org/news-releases/893024
10. Rolleston, W. (2019, May 5). Changing GM policy will be good for the environment and Carbon zero *Stuff NZ Farmer.co.nz*, https://www.stuff.co.nz/business/farming/opinion/112596782/changing-gm-policy-will-be-good-for-the-environment
11. VIB. (2019, April 15). Permit for CRISPR maize field trial that aims to measure climate stress. *VIB News*, https://vib.be/news/permit-crispr-maize-field-trial-aims-measure-climate-stress
12. Genius. (2018). *The project*. https://www6.inrae.fr/genius-project/Communication/Point-on-the-Project-GENIUS-as-of-July-31-2018
13. Michalopoulos, S. (2019, December 10). Corteva signs first major gene editing deal with European company. *EURACTIV.com*, https://www.euractiv.com/section/agriculture-food/news/corteva-signs-first-major-gene-editing-deal-with-european-company
14. Dobrovidova, O. (2019). Russia joins in global gene-editing bonanza. *Nature, 569*, 319–320. https://doi.org/10.1038/d41586-019-01519-6
15. Yissum. (Access 2021 & 2023). *Overview,* http://www.yissum.co.il/

16. Kumar Bari, V., Abu Nassar, J., Marzouk Kheredin, S., Gal-On, A., Ron, M., Britt, A., Steele, D., Yoder, J., & Radi, A. (2019). CRISPR/Cas9-mediated mutagenesis of carotenoids cleavage di-oxygenase in tomato provides resistance against the parasitic weed. *Phelipanche Aegyptiaca Scientific Reports, 9,* 11438.
17. Karembu, M. (2021). *Genome editing in Africa's agriculture 2021: An early take-off.* International Service for the Acquisition of Agri-biotech Applications (ISAAA AfriCenter), Nairobi Kenya.
18. Opoku Gkapo, J. (2013, June 13). *Gene editing could save Ghana's cocoa from extinction, scientists say Alliance for science,* https://allianceforscience.cornell.edu/blog/2019/06/gene-editing-save-ghanas-cocoa-extinction-scientists-say/
19. Ma, A. (2020). *Biotechnology in China: A status report* (p. 64). Foundation for Policy Innovation.
20. Penn State University. (2021, August 4). Cocoa CRISPR: Gene editing shows promise for improving the 'chocolate tree'. *Penn State News,* https://news.psu.edu/story/&521154/2018/05/09/research/cocoa-crispr-gene-editing-shows-promise-improving-chocolate-tree
21. Ricroch, A. (2020). Les biotechnologies en Chine: investissement stratégique et massif dans l'édition du génome. *Monde Chinois, Nouvelle Asie, 61,* 54–69.
22. Krinke, C. (2023, February 14). OGM: la Chine fait un pas vers les Etats-Unis. *Info'OGM veille citoyenne,* https://www.infogm.org/7664-ogm-chine-fait-un-pas-vers-etats-unis
23. Toppan, A. (2021). Au Panthéon des Plantes modèles. In C. Regnault-Roger (dir), *La culture du tabac en France, sauvegarder un savoir-faire, promouvoir l'innovation?* (pp. 225–242). Presses des Mines.
24. Regnault-Roger, C. (2020). *Des plantes biotech au service de la santé du végétal et de l'environnement* (p. 56). Fondation pour l'innovation politique.
25. Calyxt. (2020, April 7). *Calyxt Licenses New Enabling Technology from University of Minnesota for Greater Efficiency in Gene Edited Plants,* https://calyxt.com/calyxt-licenses-new-enabling-technology-from-university-of-minnesota-for-greater-efficiency-in-gene-edited-plants/
26. European Commission. (2021, April 29). *EC study on new genomic techniques.* Letter to the Portuguese presidency, https://ec.europa.eu/food/plants/genetically-modified-organisms/new-techniques-biotechnology/ec-study-new-genomic-techniques_en
27. SEMAE. (2021) *Études et données statistiques statistical studies and data,* https://www.gnis.fr/etudes-donnees-statistiques-semences/
28. UFS. (2021, March 8). *Les nouvelles techniques de sélection végétale en 2021: avantages, limites, acceptabilité.* Contribution écrite Audition publique de l'OPECST, https://www.vie-publique.fr/report/281292-report-on-new-plant-breeding-techniques-in-2021

Part IV
General Conclusion

Chapter 17
The European Union at the Crossroads of Biotechnological Pathways for the Future

Biotechnological Innovation: A Challenge for Agri-Food and Medical Sovereignty

Our species, *Homo sapiens sapiens*, freed itself from the hazards of gathering and hunting by creating agriculture. The tools humans invented have made them evolve. Their thought has asserted itself, opening new fields of development to their intelligence. Innovation is at the heart of this evolution and has allowed them to face the changes that novelties provoke in order to better keep the advantages and limit the unwanted effects, inconveniences, dangers, and risks.

In doing so, human population has grown from a few million to 7.7 billion in just a few centuries: men, women, and children who must be fed and to whom our humanity, in its essence, must ensure decent living conditions for all, in the cities and the countryside, in the tropics, or in the taiga. The notion of progress is nowadays vilified. The finiteness of the planet frightens Cassandras who invoke overpopulation and advocate degrowth. But do we want to return to the age of the candle when we are not able to do without our cell phone?

Biotechnology is one of the new tools that the twentieth century has developed. These technologies applied to living organisms reproduce natural phenomena in the laboratory, freeing themselves from the uncertainty of spontaneous modifications of the genome that occur in nature.

These tools have evolved over the years. Jennifer Doudna, co-inventor with Emmanuelle Charpentier in 2012 of the CRISPR/Cas9 technique that creates a real breakthrough in genomic modifications performed in the laboratory, presented the advances made in recent years:

© The Author(s), under exclusive license to Springer Nature Switzerland AG 2023
C. Regnault-Roger, *Biotech Challenges*,
https://doi.org/10.1007/978-3-031-38237-6_17

All the technologies in the past were sort of like sledgehammers, but now we're basically able to have a molecular scalpel for genomes.[1]

Should we refuse the progress of being more precise, of modifying only what needs to be modified without undesirable effects in order to obtain the desired result? Should we give up improving?

First-generation biotechnologies are therefore less sophisticated techniques, but they have saved time in the development of the desired modifications. Thanks to genetically modified microorganisms (bacteria, viruses), human hormones manufactured in fermenters and vaccines that can better treat or prevent diseases such as diabetes or hepatitis B have been developed. In agriculture, varietal improvements have been made so that cultivated plants can better protect themselves against insect pests or diseases and so that weeding work is less painful for the farmer. Despite their imperfections, these first-generation techniques have led to decisive therapeutic advances that have enabled millions of patients to better cope with their illnesses and many poor farmers in developing and emerging countries (17 million in 2019) to reduce the need for phytosanitary treatments on their plots while increasing crop yields and their incomes, as the example of Bt eggplant cultivation in Bangladesh shows. These are not the only results of the implementation of these first-generation biotechnologies, but for that reason alone, the approach was worth trying.

The societal controversy on GMOs was created from scratch by NGOs that are enemies of progress. Contrary to its nice name ("a green peace"), the NGO *Greenpeace*, at the forefront, is multiplying black actions against innovative technologies, including biotechnologies. Several authors, journalists, scientists, and academics have underlined in their works[2] that the defiance toward GMOs has been organized not on scientific grounds but on those of political choices, of social choices. A minority wants to impose its conception of development by striking actions, sometimes symbolic, often destructive (ransacking of farmers' fields or experimental field trials, vandalism in laboratories or warehouses, etc.). In addition to this, there has been a regulatory and parliamentary guerrilla warfare for several years.

These exactions, which benefit from a great judicial leniency in France, have been encouraged by the complacency, even complicity, of the political world. In the process of the 2007 *Grenelle de l'environnement*, an agreement was made between the President of the Republic, Nicolas Sarkozy, and the elected representatives of the political ecology movement (today's EELV party, *Europe Écologie Les Verts*),

---

[1] Kevin Loria, the researcher behind "the biggest biotech discovery of the century" found it by accident.   Insider   July   7,   2015,   https://www.businessinsider.com/the-people-who-discovered-the-most-powerful-genetic-engineering-tool-we-know-found-it-by-accident-2015-6?r=US&IR=T

[2] For example, Hervé Kempf (2003), *La Guerre secrète des OGM*, Seuil; Jean-Paul Oury (2006) *La querelle des OGM*, PUF; Jean-Paul Jaillette (2009) *Sauvez les OGM*, Hachette; Gil Rivière-Wekstein (2012) *Faucheurs de science*, Le Publieur; Marcel Kuntz (2014) *OGM, la question politique*, PUG; Bernard Le Buanec (2016) *Les OGM: pourquoi la France n'en cultive plus?*, Presses des Mines.

to which the Prime Minister of the French government in office that year testified in a book[3]. This arrangement consisted of sacrificing GMO crops on the national territory in exchange for neutrality toward nuclear power: a fool's bargain since the Fessenheim nuclear power plant was closed in 2020 (President Emmanuel Macron) due, again, to political agreements.

Since 2007, in France and in the European Union, there has been, at best, a lack of understanding of the biotechnological stakes, at worst a certain cowardice and resignation in the face of environmental activists, and even connivance, which has led to the bending of European regulations on GMOs in order to facilitate the rejection of this technology. Today, agricultural GMOs are only cultivated in the EU in Spain and Portugal, countries that, like the Gallic Village of Asterix and Obelix, resist!

But the situation will change in the next few years. Indeed, second-generation biotechnologies (NGTs), as this book has pointed out, hold out even greater hope for agriculture, human, and veterinary medicine and animal welfare. If a small number of products derived from NGTs are currently commercialized (in the field of food biofortification and nutraceuticals), there are many applications at the R&D stage that will allow for a market launch within 5–10 years.

The patents taken on these technological advances will be decisive for the agri-food or medical independence of countries.

There is a global race to file patents, with two champions who are on an equal footing: China and the United States. China has wisely developed a research policy by sending students to train in leading laboratories (the United States, European Union, the United Kingdom, Canada, etc.) and then returning to the country to advance research in the field. It has also encouraged the takeover of technologically advanced companies, such as the Swiss company Syngenta by the state-owned consortium *ChemChina*, to create synergies that allow it to be a leading country in the sector.

France, but also the European Union, is not very dynamic in this sector. It is true that research investments are made through European framework programs for research and technological development (FP7 (2007–2013), then H2020 (2014–2020), and now Horizon Europe (2021–2027), but the results obtained are implemented elsewhere in the world. The fault lies in administrative red tape and especially in European regulations that are disproportionate to the risks involved.

The Court of Justice of the European Union (CJEU) has ruled that the GMO regulations initially defined in 1989–1990 and consolidated in 2001 by Directive 2001/18/EC should be applied to second-generation biotechnology products. This directive, which was conceivable in 2001, has now become obsolete due to the progress in scientific knowledge that has occurred over the last 20 years. The decision of the CJEU puts the European Union at odds, not only in terms of technological development but also in economic terms of circulation of goods in a globalized market. Do we want, as with GMOs, to refuse the cultivation of genetically modified varieties (some of which are undetectable in the finished product) in order to

---

[3] François Fillon (2015) *Faire*, Albin Michel.

better import them and thus follow in the footsteps of countries which, moreover, do not differentiate, today, in their shipments, between GMO and conventional soybeans or corn that arrive mixed together?

Many countries have seized on genome-editing techniques to project their development, even when their means are limited. Collaborations between international and American institutions and African laboratories have been established to enable this continent to have answers adapted to local conditions and its environment.

The rejection of these biotechnological advances will inevitably lead to the economic decline of the EU. The fact that Calyxt, a subsidiary of the French company Cellectis (originally a start-up of the Pasteur Institute), had to become Americanized to continue its growth speaks volumes about the obstacles that exist in our country to allow promising economic development.

Fortunately, the European Commission has opened a door in April 2021 for a revision of the regulations applied to plant applications, treating health applications separately. A consultation is underway.

The *French Association of Plant Biotechnology* (AFBV) recently remarked, in a press release at the 2022 Paris International Agricultural Show, that "Europe must urgently take hold of new biotechnologies to safeguard its food and economic sovereignty" by adapting the current regulations on GMOs, if only to limit production losses in the fields. This objective is part of the virtuous chain of the fight against food waste in order to face the increase of 10 billion inhabitants by 2020.

But not only agricultural plant applications should be concerned. Animal applications (veterinary medicine and animal welfare) are also worth considering. The example of the *Acceligen* Company in Minnesota (United States) highlights the dynamism of this sector.

We hope that this book, which gives a panorama of what is being done in the field at a given moment, will show the urgency of modifying European regulations on GMOs and NGTs, of adopting a confident attitude toward technological progress, and of marginalizing the precautionary principle which hinders the dynamism of development in France and in the EU by becoming, at best, a principle of inaction and, at worst, a principle of retrogression, because whoever does not progress necessarily regresses.

At the Heart of Scientific and Academic Reflection

As you can see, this book is not only an author's book but also a cry for *fake news to* stop clouding the scientific debate and for society, through the example of biotechnologies, to be reconciled with the scientific world.

It is by joining the Scientific Committee of the *High Council for Biotechnology* (HCB), on which I sat for its entire duration, from 2009 to 2021, that I have measured the extent to which the divisions on this key issue for our future are caused by certain experts who are above all activists guided by a political ideology and are free from scientific rigor.

I dedicate this book to the large majority of members of the HCB Scientific Committee who have been committed to the scientific approach throughout their mandate, with a special mention to Jean-Christophe Pagès who chaired this committee and then the whole HCB during these 12 years, keeping the scientific spirit accompanied by a benevolent courtesy.

I also salute the memory of Louis-Marie Houdebine who left us on March 1, 2022. A French researcher (research director at INRA) and pioneer in animal biotechnology, member of the *French Academy of Agriculture* since 2003 and also president of the *French Association for Scientific Information* (AFIS) between 2011 and 2014, he worked all his life to enlighten the scientific community and the general public with his knowledge and to counter misinformation on GMOs.

I also do not forget the rich debates of the working group *Genetically Modified Plants* of the *French Academy of Agriculture* led by Jean-Claude Pernollet. This work resulted in a collective book entitled *Plantes génétiquement modifiées: menace ou espoir?* [Geneticcally Modified Plants : Threat or Hope?] (2015, Quae). Another bi-academic working group, Académie des technologies-Académie d'agriculture de France, led by Bernard Le Buanec, has examined the technical challenges facing agriculture and the responses provided by technologies, among which biotechnologies occupy a prominent place. A collective work was published as a result of this work in 2019 (*L'agriculture face à ses défis techniques, l'apport des technologies* [The challenges facing agriculture technical challenges, the contribution of technologies], directed by Bernard Le Buanec, Presses des Mines, collection *Académie d'agriculture de France*). Other working groups have provided reflection on the subject, led by Agnès Ricroch on green biotechnologies (2014–2017) and by Paul Vialle and Bertrand Hervieu on the ethical aspects and societal acceptability of rewriting the genome (2016–2019), not to mention the work of the sections, including the Plant Production Section and the Agronomists and Geneticists Section (which welcomes me, even though I am neither). The Academies are cenacles of erudition and reflection.

Not to be outdone, the Academy of Sciences jointly organized with the French Academies of Technology and Agriculture a memorable interacademic session on *Genetically Modified Plants* on November 19, 2013, at the *Institut de France.*

I also remember the sessions of the *National Academy of Pharmacy* during which its Permanent Secretary, Agnès Artiges, invited me to give presentations on the topic of biotechnologies and the discussions that followed.

A Worldwide Recognition

While Paul Berg received a shared Nobel Prize in 1980 for his work on nucleic acids, it took 30 years for the scientists behind the first transgenic transformed plants to receive a joint award for this innovation.

Marc Van Montagu, Mary-Dell Chilton, and Robert Fraley, described by the press as "GMO apostles,"[4] were indeed joint winners of the *World Food Prize Foundation* in 2013. This prize, considered by some to be the Nobel Prize for food, is defined as the most prestigious international award that "honored outstanding individuals who have made vital contributions to improving the quality, quantity, or availability of food throughout the world."[5]

Feeding and caring for people are indeed at the heart of the challenges of our future.

The Nobel Prize in Chemistry was awarded in 2020 to Emmanuelle Charpentier and Jennifer Doudna, just 8 years after their discovery of CRISPR/Cas9, which propelled genome editing into the future. Should we see in this diligent recognition that we have finally understood the progress that biotechnologies bring to humanity?

---

[4] Claude-Marie Vadrot, *Politis,* June 27, 2013.

[5] https://www.worldfoodprize.org/en/about_the_foundation/

Presentation of the Authors

Catherine Regnault-Roger is a pharmacist from René Descartes University (Paris University) and doctor of State in natural sciences from Pierre and Marie Curie University (Paris Sorbonne).

She has had an academic career in France (University Denis Diderot-Paris VII, CNAM Paris, and University of Pau and Pays de l'Adour [UPPA]) and abroad (French cooperation at the University of Constantine in Algeria and at the Faculty of Pharmacy and Dentistry of Monastir in Tunisia), during which she has climbed all the academic ranks and grades (from assistant professor to professor of exceptional class). She is now an emeritus professor at the *University of Pau and Pays de l'Adour (E2S-CNRS)*. She has created within UPPA several university diplomas, at Master level on biotechnologies and plant bioprotection, at graduate level to encourage student mobility abroad (DUEI: *Diplôme Universitaire Études Internationales*) or in companies (DUFAST: *Diplôme Universitaire Formation Appliquée en Sciences et Techniques*).

At the research level, she is recognized for her work on the bioprotection of agrosystems and the environment through an approach of chemical ecology and sanitary quality of crops (mycotoxins) and then for transdisciplinary research on biotechnologies, biocontrol, and agricultural revolutions. She has served on numerous university committees, from the *National Council of Universities* (2003–2013) to the *Biological Monitoring Committee* (2010–2015) of the *Ministry of Agriculture* and the *National Commission on Engineering and Expertise* (CGRA1) of the *Institute for Research and Development* (IRD) (2003–2007) as vice president in both instances.

Member of the Scientific Committee of the *High Council for Biotechnology* for its entire duration (2009–2021) as an expert in ecotoxicology and regulatory issues on post-marketing surveillance of transgenic crops, she is also an emeritus member of the *French National Academy of Pharmacy* (Environmental Health Section) and a full member of the *French Academy of Agriculture* where, after having been Secretary of the Plant Production Section, she is in charge of the *Books Committee*

C. Regnault-Roger, *Biotech Challenges*, https://doi.org/10.1007/978-3-031-38237-6

of the Academy and director of the *Académie d'agriculture de France* collection published by Presses des Mines. She is also a member of the Editorial Board of the journal *Phytoma – la santé du végétal* (Groupe de la France agricole) and of the *Editorial Advisory Board* of the journal *Industrial Crops and Products* (Elsevier). She has written and coordinated several scientific reference books (in French, English, Spanish) which have received awards as well as book chapters and numerous scientific and technical articles. The *French Association of Plant Protection* awarded her its Medal of Excellence in 2015. She is a knight in the Order of the *Legion of Honor* and an officer in the Order of *Agricultural Merit*.

Foreword for French Edition: Jean-Yves Le Déaut

He holds a PhD from the *University Louis Pasteur* in Strasbourg (1976) where he was an assistant in the Faculty of Medicine (1969–1976) with a stay at the Faculty of Sciences in Tunis (1971–1973) as a volunteer for active national service abroad. He was a lecturer and then a professor at the University of Tananarive and a researcher at the *Pasteur Institute of Madagascar* (1977–1983) as a cooperant. Back in France, in Nancy in 1983, he became professor of biochemistry at the *University Henri Poincaré* (1983–1998). He was also a lecturer at *Sciences Po Paris* on "*Les grands enjeux du XXIè siècle*" [The Major Challenges of the 21st Century] from 2006 to 2013.

Jean-Yves Le Déaut became a politician after having been elected deputy in 1986 in a Meurthe-et-Moselle constituency. He also held several local mandates in Pont-à-Mousson and Nancy and in the Lorraine Region where he was elected as first vice president from 2004 to 2013 in parallel with his national mandates as PS (Socialist Party) deputy, reelected for 31 years. He is today honorary member of Parliament.

He chaired the Parliamentary Office for the Evaluation of Scientific and Technological Choices (OPECST) several times between 1989 and 2017 and was chairman of the information mission on *Climate Change: The Major Challenge* (2006), member of the Parliamentary Assembly of the Council of Europe (2012–2017), general rapporteur for Science and Technology (2013–2016), and member of the *Strategic Council for Research in France* (2013–2017). He also chaired the *National Assembly's Information Mission on GMOs* (2008 law). He is the author of several parliamentary reports including the report on "*The Innovation Principle*" (OPECST 2014), the report "*The Place of Massive Data Processing (Big Data) in Agriculture*" (OPECST, July 2015), the report on "*Plant Genetic Resources from Improvement to Conservation of Species: The French Model*" (OPECST 2016), the report "*From Biomass to the Bioeconomy: A Strategy for France*" (OPECST 2016), the report "*Innovation and Climate Change: The Contribution of Scientific and Technological Assessment*" (OPECST 2015), the report "*The Evaluation of the National Research Strategy 'France Europe 2020'*" (OPECST 2017), and finally the report "*The Revolution of Targeted Genome Modification (Genome Editing)*

Economic, Environmental, Health and Ethical Stakes of Biotechnologies in the Light of New Research Avenues" (OPECST 2017). He is a Knight in the Order of the *Legion of Honor* and a Knight in the Order of the *Palmes académiques*.

Foreword for English Edition: Marc Van Montagu

Marc Van Montagu was born in Ghent (Belgium) and studied at the University of Ghent until his PhD in chemistry/biochemistry in 1965. He joined the Department of Cell Biology at the Medical School of Ghent University and focused his research on bacterial RNA. He then worked with Jeff Schell (1935–2003) on crown gall disease. In 1974, they discovered the mechanism of transfer of the disease to the plant by a "Ti plasmid" of the soil bacterium *Agrobacterium tumefaciens*. They developed the first genetic transfers by transgenesis and created the first transgenic tobacco capable of protecting itself by secreting insecticidal proteins.

He is a university professor and director of the genetics laboratory at *Ghent University* and was also scientific director of the genetics department of the *Flemish Institute for Biotechnology* (VIB). Together with Jozef Schell, he founded the company *Plant Genetic Systems Inc.* in 1982 and set up the company *CropDesign*, which develops biotechnological agronomic varieties. Later, he chaired the *European Federation of Biotechnology*, a nonprofit association for the promotion of biotechnology. Winner of prestigious prizes (*Japan* Prize in 1998, *World Food Prize Foundation* in 2013) and holder of numerous *honorary* doctorates from European universities, he was knighted by King Baudouin of Belgium in 1990. He is a member of several scientific academies including the *National Academy of Sciences of America*, the *Russian Academy of Sciences*, the *Academia Europaea*, the *Royal Academy of Overseas Sciences*, and the *French Academy of Agriculture*.